Being a Researcher

Nollaig Frost

Open University Press

Open University Press
McGraw Hill
Unit 4
Foundation Park
Roxborough Way
Maidenhead
SL6 3UD

email: emea_uk_ireland@mheducation.com
world wide web: www.mheducation.co.uk

Commissioning Editor: Beatriz Lopez
Editorial Assistant: Sarah Moorhouse
Content Product Manager: Graham Jones

A catalogue record of this book is available from the British Library

ISBN-13: 9780335251605
ISBN-10: 0335251609
eISBN-13: 9780335251612

Typeset by Transforma Pvt. Ltd., Chennai, India

Fictitious names of companies, products, people, characters and/or data that may be
used herein (in case studies or in examples) are not intended to represent any real
individual, company, product or event.

Printed and bound by CPI Group (UK) Ltd, Croydon, CR0 4YY

Praise page

Being a Researcher *offers an engaging and enlightening journey through the personal complexities of doing research. It examines what 'good research practice' means in the context of how different aspects of the self (such as personal motivations, ethical principles, interpersonal skills) can enable researchers to build on their strengths to produce their very best work. This accessible and original book is essential reading for all researchers, regardless of their academic discipline, methodological preferences or career stage.*

Professor Amanda Holt, Professor of Criminology,
University of Roehampton, London

Sometimes reading a book makes 'everything' slot into place. It helps you to look at reality from new perspectives with a 'map' that makes utter sense. This book approaches research from a wonderfully human - but often neglected, angle.

What is a 'researcher'? Why do people become researchers? These questions make starting points for a well-written, insightful and knowledgeable exploration based on the author's rich experience. This book brings 'doing research' to a real, human level - dispelling the myth, once and for all, about research being something 'out there', detached from those doing it. It will likely be a go-to for anyone considering research. In an inspiring, well-structured and easy-to-follow way, Professor Frost takes us on a journey "that starts by considering what a researcher is, continues to think through the characteristics of researchers and culminates in equipping you to understand what it means for you to be a researcher". I expect to see this book as recommended reading on courses ranging from clinical Diploma to Doctoral level training; I cannot recommend it enough.

Dr Sofie Bager-Charleson, the Metanoia Institute,
author of *Enjoying Research in Counselling and Psychotherapy*

Many books discuss doing research, but few address being a researcher. This book fills that void, and does it well, offering considered insights into what it means to be, and what is involved in being, a researcher. The book draws from the author's extensive experience in doing, teaching, and supervising research, weaving in relevant scholarship to sustain its arguments. The coverage is comprehensive, offering thoughtful discussion of researcher values, roles, practices, relationships, feelings, and approaches, effectively capturing the full breadth and depth of being a researcher. The writing is always clear and focussed, and the book well structured. This will be a invaluable book for those entering the research field, and a valuable resource for those already there.

Kerry Chamberlain, Professor Emeritus,
School of Psychology, Massey University, Auckland, New Zealand

Contents

About the Author

Nollaig Frost is an Independent Academic, Adjunct Professor at University College Cork, Ireland, Senior Lecturer at Metanoia Institute, and Visiting Lecturer at City St George's, University of London. Her keen interest in who researchers are has arisen from her own research practice into motherhood and from many years of teaching and supervising research students. She is the author of *Qualitative Research Methods in Psychology: Combining core approaches (2e)* and *Practising Research: Why you are always part of the research process even when you think you're not.* She has been a member of the European Association for Qualitative Researchers in Psychology since its founding in 2017 and, most recently, has taken up the role of Editor-in-Chief of the *European Journal for Qualitative Research in Psychotherapy.*

Acknowledgements

This book has arisen from many years of not only doing my own research but of teaching and supervising research students at all levels and in many contexts. That work has given me much food for thought, insight into how researchers experience and understand the research process and plenty of enjoyment. I wish to thank all the students I have encountered and to express my pride in the researchers many of them have become.

Alongside the students, there have been colleagues who are researchers. The collegiality, generosity and encouragement from people like Stephen Frosh, the late Amal Treacher, Sofie Bager-Charleson, Carla Willig, John McCarthy, Toni Bifulco, Annabelle Mark and Linda Finlay have been inspirational as well as educational. Others such as Maria Dempsey, Sarah Foley, Sarah Robinson, Amy Lucas, Angela Veale and Deborah Bailey-Rodriguez have been constant supportive presences over many years, and without their contributions, this book would be poorer. To my colleagues in the European Association for Qualitative Researchers in Psychology (EQuiP), I say thank you too, for enabling me to learn about and discuss being a researcher in different international contexts, whilst also having a lot of fun. To the many other colleagues whom I have met and worked with at Middlesex University, University College Cork and City St George's, University of London, amongst others, I extend my gratitude for being so influential on my learning about researchers. To all my current students – keep going, you will get there!

Many thanks to the editors at Open University Press. Led by Laura Pacey, they have been understanding of the interference of life in the writing of books and shown only compassion and support over the delays.

Of course, as ever, I thank my family for their tolerance of me and my book. Nick, Sam and Leona – thank you, and apologies.

And, finally, to my wonderful community in West Cork, thank you for putting up with me boring you with the progress of the writing and for teaching me more than any book could.

This book is dedicated to my late parents, Derek and Eleanor, who only ever showed me pride and encouragement, and to the late Jerry Sheehy and DJ Courtney for educating me in how to keep learning new skills.

Part **1**

Who Researchers Are

1 Becoming a Researcher

Introduction

This book is about what it is to be a researcher. Its premise is that *being a researcher* is different to *doing research*. Being a researcher means that you are curious to find out and understand more about aspects and problems of the world, and to see how they can be addressed and solved. In one sense, we are all researchers all the time – we explore, hypothesize, ask questions and seek explanations, so that we can make sense of behaviours, problems and experiences. However, to be a 'formal' researcher – academic, practitioner or community-based – means going beyond ideas, opinions, beliefs and hunches to build a trustworthy and reliable knowledge base that can be shared confidently with others. It means carrying out research that is formalized, rigorous and based on techniques and approaches that are valued and recognized. Being a formal researcher requires skills, aptitude and resources that can be used with your own curiosity to conduct good research. To be a successful researcher – and one who enjoys doing research – therefore, means recognizing personal characteristics and understanding how to nurture them to enhance your research practice.

Often the path to becoming a researcher begins with having to carry out research as part of study or training requirements. For some, conducting research for the first time is daunting and can feel like a burden. For others, it is exciting to investigate a topic they are passionately interested in. For most, the start of research experience lies somewhere in between, becoming less daunting as skills and knowledge grow, or less attractive as the interest gives way to drudgery. This book aims to take you on a journey, from wherever you are in your research experience, that starts by considering what a researcher is, continues to think through the characteristics of researchers and culminates in equipping you to understand what it means for you to be a researcher.

So, this book is not about doing the research as such; it is more about the person doing the research. Who does research and who does it well? What is it about you that will help you to develop an interest in research or realize that it is not for you? Why does your fellow researcher love it so much when you don't? What can you do to ensure that, whatever the reason you are doing research, you will do it to your best ability and ensure it is of the highest standard?

The aim of the book is to support and inform you as a researcher, wherever you are in your research career, whatever research approach you are using and whichever discipline you are pursuing. There is reference throughout of important researcher considerations such as objectivity/subjectivity, positionality and reflexivity in different contexts and from different perspectives.

Whilst at times these may seem to be more salient to researchers in the social sciences and humanities, I believe it is also the case that all researchers should have awareness of these and their influence, so that choices can be made about their approaches and stances as researchers. The book will show you which aspects of your self are important to being a researcher and how recognition of who you are and what you bring to your research can help develop both the quality and enjoyment of your research practice. It does this by emphasizing that the techniques of research practice only form part of what is necessary for good research. It highlights the role and value of researcher motivation and shows that whilst there are tasks and systematic approaches that must be used, research quality is elevated by the understanding that it is not only what a researcher does that is important but also who a researcher is. Alongside practical considerations about your time, incentive and resources, the book will consider what is required of you as a person, as well as the relationships you will form and how to manage them as you carry out research. It will reflect how to integrate research into your study, practice and life. Along the way, challenges, tensions and barriers to researchers will be examined and useful tips provided. Key issues such as self-reflection and positionality are returned to throughout to consider and understand them from different perspectives. The book draws on reflections from researchers with different levels of experience to illustrate what can be learnt from others and discusses different research environments and settings. Overall, the book aims to help you to identify useful attributes and settings, and practical considerations for researchers, whatever the reason they are doing research and wherever they are doing it.

I hope it will help you to think about who you are as a researcher, what you share with other researchers and how you differ from them. Importantly, I hope it will help you question and understand what being a researcher means for you.

Structure of the book

This book is divided into four themed parts. Each part includes three chapters, each of which focus on a different aspect of the overarching theme. The parts fit together to take you on a journey through what it is to be a researcher.

Part 1: *Who Researchers Are*: a look at why people do research, what is required of a researcher to be successful and what makes research enjoyable or not for researchers.

Part 2: *Researchers in Different Settings*: a focused consideration of three key research settings – academic, practice-based and community-based. Of course, there are overlaps but each setting also has its own expectations, demands and challenges.

Part 3: *Researcher Choices*: a look at issues such as knowing your role and influence in different research approaches and addressing challenges to them, working with different forms of data and understanding how they relate to personal characteristics, and researching with people and communities you may or may not have knowledge of or things in common with.

Part 4: *Researcher Identities*: this part considers how researchers put theory into practice and what inhibits or helps them in this, emotions generated or provoked by doing research and what it is like to be researched and assessed as a researcher.

What is a 'researcher'?

As human beings, we are curious about the world around us and research it by seeking out information about it, testing out hunches and forming explanations that help us make sense of it. A researcher is 'someone who studies a subject, especially in order to discover new information or reach a new understanding' (dictionary.cambridge.org). Combining these enables us to arm ourselves with existing and new knowledge to choose (and change) our behaviours, inform our perceptions and form an understanding of who we are and what it is to inhabit our world as individuals. In this book, we take these strategies as starting points for understanding what a formal researcher is and we use them to consider how this approach to life contributes to how we carry out organized research in academic and other settings. A researcher is, therefore, a person who carries out rigorous investigation into a topic of interest in ways that go beyond opinion and ideas, so that it is testable, rigorous and trustworthy for a wider audience.

Why do people become researchers?

People become researchers for different reasons. Often the starting point is the requirement to conduct research as part of studying for academic or professional qualifications. Some then want to use these skills in their professional roles, to address concerns they have about their community or to make a career out of doing only research.

There are often personal reasons too for people becoming researchers; they may wish to address burning questions that have arisen in their personal or professional lives, hone problem-solving skills, work more closely with colleagues and mentors, learn more about working as part of a team or develop team leadership experience. Practical benefits to becoming a researcher may be attractive too: developing transferable skills such as time and budget management, learning to use online tools and acquiring or extending communication and presentation skills. Being a successful researcher is a good way of showing colleagues and potential employers that you can complete work and multitask. Writing is a key part of being a researcher and many enjoy the challenges of presenting research in a way that is meaningful and accessible to others. And let's not forget the clear advantages of useful additions to CVs in the form of published articles, research reports and conference presentations.

People, therefore, become researchers by design, opportunity or requirement. They may do so out of necessity, to satisfy their curiosity or as a way of developing a fulfilling career. The time in which they are a researcher may be short-lived or life-long and be the entire focus of what they do or complement other roles. Everyone who becomes a researcher, though, shares a curiosity about the world, a desire to understand more about it and an ambition to understand it better. The following reflection highlights the value of being a researcher for someone who is also a teacher.

Researcher reflection (from an experienced academic researcher)

I am not only a researcher, but without being a researcher, many other aspects of my job would be less interesting. Being a researcher helps me to explore and understand more about people, what happens to them, and what their experiences mean to them. I can use the knowledge I generate through my research to develop my teaching practice, enable me to support students, and to develop my confidence in my role. In turn I can use my experience in my non-researcher roles to form new research ideas. If I was not a researcher, I think I would feel like I am less grounded in what I know, what I think I know, and what I know I don't know.

Researcher characteristics and qualities

There is a myriad of characteristics that researchers need (such as a desire to find answers); a number of these will enhance their enjoyment and success in research (such as curiosity, interest and passion) whilst some are requirements (such as critical thinking and commitment). Researchers need to have knowledge and skills to know what research is and how to do it. They need to be able to manage their time, overcome challenges and be flexible in their thinking. They need to be able to work alone and with other people, to be collegiate team members and supportive team leaders. Researchers need to be able to sit at a desk for hours, and to be able to go out into the field or laboratory to meet with people. They need to be able to read papers and books relevant to their topic and to write reports and articles. Researchers need to be able to get projects off the ground and see them through to the end. Some of these qualities may be innate to a person whilst there might be others that can be learnt, but there are some that can make a real difference between doing research and *enjoying* doing research. In the section below, we consider some personal characteristics and qualities that can be important to enjoying being a researcher.

Curiosity

This is top of the list for enjoying research. Curiosity means wanting to know about things, finding out more and developing understanding and insight. To be curious means to pursue, to explore and to test an interest in a topic. A researcher who has curiosity is likely to be innovative in their thinking about research yet rigorous about making meaningful investigations. They will enjoy the adventure of research and relishing reaching new insights. They will get a thrill from being the person who has found something that hasn't been previously known, and will know that part of the challenge is to find ways over and around apparent barriers on the way.

Passion

Researchers know that pursuing an investigation can be a long process of advances and setbacks. Keeping the focus of the research can be hard but if the topic is one that the researcher is not only interested in but has a passion for, they can relish the journey of understanding more about it. Passion for researching a topic may come from burning questions, personal experience, observation or anecdote; however it originates, being passionate about what is being researched will sustain a researcher throughout the research process.

Communication

The stereotypical picture of a researcher is commonly of a person in a white coat, looking serious and working alone, usually in a laboratory. In practice, however, being a good researcher often means working with others in various settings and in many ways. To be good at communicating, and to enjoy doing so, will help greatly in progressing the research by enabling problem-sharing and promoting critical thinking. Effective communication will also help to maximize full involvement from all those involved in the research (participants, stakeholders and co-researchers) because they will feel included, informed and valued. Being a researcher who is a good communicator means new ideas can be generated and discussed, knowledge relayed and shared, and questions raised and addressed. The communication is likely to be in verbal, visual or written form, depending on its purpose and audience. Having skills in communicating in different ways is important for researchers. Whatever the purpose or form of communication, researchers who appreciate other people and want to bring them on the journey with them will be researchers who enjoy what they are doing.

Integrity and honesty

A good researcher is, of course, ethical in their research practice. A researcher who enjoys their research is one who seeks to ensure the highest level of honesty and transparency in it because they want the research to be meaningful to and impactful for as many people as possible. Having these characteristics will surmount challenges and competition, to ensure that the process from start to finish is one that they can be proud of and will produce results they take pride in standing by. The achievement of adding trustworthy and respected knowledge to the field can bring an additional layer of enjoyment to the research.

Resilience

Whilst many researchers enjoy many aspects of the research process, there are undoubtably setbacks and upsets in almost every study; recruitment may be hard, results not what were expected, data not rich enough or papers not accepted by journals – the list goes on. Furthermore, life often intervenes, causing starts and stops in the research so that you can attend to personal situations and relationships, other tasks with tighter deadlines or a need to find new methods or other resources to complete the research. The researcher who can withstand these issues and work with them without becoming too frustrated or worn down by them is the one who will enjoy their research more. Knowing when to take a break from research is an important aspect of being resilient to its challenges. Similarly, being flexible in addressing challenges and accepting of compromise in the name of producing good research helps in remaining committed to it. Enjoyment of the process comes not only from reaching results but also from navigating the challenges that arise in the path leading to them.

Open-mindedness

Being open-minded as a researcher means not only that you minimize the likelihood of biasing your research but also that you are more likely to be able to think through problems and challenges in more creative ways. This may enable you to be more consultative with colleagues so that ways of addressing barriers to research progress can be drawn from other fields or disciplines, or it may mean that when research is not going (or has not gone) the way that you expected, you will be able to think through what is happening and why, in a more productive manner. Open-minded researchers are more likely to be able to manage contradictory or unexpected data and results. With an approach that is willing to think through possible solutions or impacts of research, the open-minded researcher is better able to advance their research in new and innovative ways.

Self-reliance

Being a researcher can often mean long hours, days or months working alone, but it can also mean collaborating with others and networking. The variety of tasks involved in conducting research brings with it excitement and a need for good time-management, goal-setting, prioritizing, meeting deadlines and working under pressure. A self-reliant researcher can plan for these and for unexpected hurdles with more confidence. Knowing what skills and resources are available and how to make best use of them brings assurance that the work will be done, and a knowledge of when and how to call on and work with others ensures that it is. The lower levels of stress and worry that this ability brings will provide the self-reliant researcher with the freedom to focus more on the research and the enjoyment of it.

> **Reflective question**: Which of the above characteristics resonate with you? What additional characteristics and qualities do you have that may help you to enjoy being a researcher?

Researcher self-care

Having some or all of the above characteristics and qualities will go a long way towards being a good researcher and one who enjoys doing research. It is also important to employ them in self-care as a researcher. There is a growing body of research that highlights the importance of researchers protecting their mental health (Hazell, Chapman, Valeix, Roberts, Niven & Berry, 2020), minimizing the risk of burnout (e.g. Daumiller & Dresel, 2020) and recognizing the potential impact of their role on their emotional well-being (Smith & Ulus 2020). Research has shown that researchers can be at risk of vicarious trauma (Berger, 2021) and of the effects of imposter syndrome (Abdelaal, 2020). The researcher carries out a wide range of tasks, in a variety of settings, with a number of people, or alone, and often under pressurizing deadlines. In addition, they may be collecting and analyzing distressing data, meeting with vulnerable people and juggling the researcher role with other roles (such as teaching, administration, professional practice or community work). All of this on top of the commitments in their personal lives. Recognition of the need for self-care is the first step towards practising it. Managing time, including taking time out when needed, are important. Connecting with others, in both social and work-related relationships, can offer support, sounding boards and opportunities to rant or celebrate. Taking up opportunities with supervisors, trainers and mentors to discuss challenges and concerns about research allows others' experience to be shared in productive ways. Similarly, the researcher making use of services that provide support to enable them to progress with minimum disadvantage is a valuable self-care strategy.

There are an increasing number of podcasts and online resources to support researcher well-being (see, e.g. https://www.editage.com/insights/top-10-well-being-podcasts-for-researchers), and of course colleagues, friends and fellow researchers can be valuable in listening to and sharing experiences. Finally, it is important not to underestimate physical well-being and its relationship with mental well-being. Sitting for extended periods at a desk, working alone and perhaps feeling isolated, and working long hours can impact how your body feels – paying attention to diet and exercise, leaving the work setting to get outside and maintaining social activities, hobbies and pastimes can all contribute to an increased sense of well-being.

Researcher relationships

Even when conducting research alone, there is a need for all researchers to have contact and connections with other people. In the early stages of being a researcher, this may be more on a learner/teacher basis but as researchers become more experienced and confident, and hopefully start to enjoy being a researcher, these relationships can be more about generating ideas, discussing challenges and clarifying questions and concerns. Given the importance of managing time when doing research, there is also value in having relationships with people outside the research environment to allow space for discussing issues that may be affecting the researcher adversely. For most researchers, there are formalized opportunities available (supervisors, trainers and mentors) and building constructive relationships with these people will aid the success and enjoyment of being a researcher as well as the quality of the research. Other relationships, less intense perhaps but equally productive, may be with researcher colleagues, others in the institution who may be responsible for administrative tasks that are important to the research, fellow students and friends in similar roles elsewhere. Of course, it is also important to attend to family relationships. Fostering and looking after personal relationships are important because they help to maintain all the other aspects of life and participation in it. Family relationships can be crucial to both the successful completion of a research study, offering support and encouragement, and acceptance that long and unsociable hours at work may need to be kept (at least for a while).

In the next sections, we will consider some of these relationships in relation to their importance to being a researcher, and provide some ideas about how best to cultivate and maintain them whilst also doing research.

Formal relationships: supervisors and mentors

The relationships with supervisors can be the most valuable and enduring of all the relationships that researchers make during their careers. A supervisor's role is sometimes compared to a parental one in which the novice researcher starts out dependent on their supervisor for support and guidance and gradually moves to a position of expertise. Bringing the supervisor along on this journey is extremely important for new researchers.

To do this effectively means recognizing the supervisor's greater knowledge and experience of doing research as well as their knowledge of the topic and/ or methodology, and being respectful of it so that the relationship is a productive and learning one. As researcher knowledge is acquired and developed, meaningful discussions and debates can be held to discuss and decide the direction of the research, the approach to it and the results from it. The developing researcher will gradually take control of these as their knowledge increases.

It is clear that power is inherent in relationships with supervisors, and it may be that this is challenging to accept for some researchers, particularly if they are experts in other fields. It can be hard to take criticism and not feel insulted if feedback suggests that writing or other aspects of your researcher role is not up to scratch. Acknowledging the supervisor's role in supporting and guiding research can help, as can working to form a relationship that has as open communication as possible to allow for discussion and questioning (Bager-Charleson, 2019).

At the same time, it is also reported that supervisors occasionally exert this power in ways that are unhelpful to researchers – not responding to emails, finding it hard to make time for meetings, not preparing for supervisory sessions and so on (e.g. Templeton, 2019). This can be stressful and anxiety-provoking for researchers. Before the relationship completely breaks down, it can be helpful to consider your own expectations of it and whether they match those of the supervisor. These can be about practical considerations around frequency and length of supervisions, what you are expected to bring to them or what feedback is reasonable to ask for. They may also be about more personal aspects, such as the ways in which you are open to receiving feedback, how you act on it or what you (and your supervisor) see as constructive feedback. It can help to consider your role and interactions in the relationship before raising concerns about them with your supervisor, as can setting agreed guidelines, and ultimately, working to preserve and develop it.

Researcher mentors are different to researcher supervisors because the role of the mentor extends across the researcher's broader professional and personal development. Mentors may be experienced in academic development, for example, and can share their networks and knowledge with you to help you progress. They can also be sounding boards on task management or when making decisions about ways forward. Overall, mentors offer insight into the directions that researchers can take in all aspects of their development.

If you are lucky enough to have both a supervisor and a mentor, then good relationships with both can help set the stage for you and your career to progress as you want.

Personal relationships

Whatever relationships you have outside of the setting in which you are a researcher, there is a need to recognize their value in providing support. Relationships with friends and family can be supportive, encouraging, (helpfully) critical and useful distractions from the intense focus of research. Having people around you who believe in what you are doing, even if they don't understand what that is and why you are doing it, can be a spur to keep going when challenges arise. They can help you keep the research and issues associated with it in perspective and can be a useful sounding board when you are juggling research with other commitments. It is important that those close to you are apprised as far as possible of the potential and actual demands of being a researcher in terms of how it might take your focus away from them for a while, cause frustrations that you may bring home or to social events, mean that you have to, or want to, miss family occasions and other events or, as when doing fieldwork, mean that you have to be away for periods of time. If people who are important to you are warned of these likelihoods in advance, and kept up to date with them as the research progresses, they are less likely to feel excluded or upset when the occasions arise. In some forms of research, such as ethnography, being a researcher can mean going to live with other people for a while, and this may cause concern to close ones. Reassurance about why this is, and for how long you may be away, can be helpful. At the same time, confidentiality must be respected, and so, those close to you may also need to be informed that there are certain details you cannot give them.

There may be demands on the time of other people close to you that means that they want your help and support in managing these demands. Whether this is childcare or a listening ear, attention to these relationships can help you as a researcher to take time away from the research and re-focus on prioritizing what is important. It may also be that these requests come at a time when you can't take time away from the research, notably when you are rushing towards tight deadlines, for example. Coming up with a compromise for those close to you will help lessen guilt and anxiety about prioritizing being a researcher.

Above all, maintaining close relationships whilst also being a researcher means that you can keep different perspectives on your life and work, and how you manage them together. This will bring enjoyment as well as valuable time to take on roles other than that of a researcher, knowing that the research will be waiting for you when you return to it.

Being a researcher in different settings

Being a researcher in different settings has different pathways and different expectations. The purposes and motivations for the research vary, and requirements of the role are informed by research areas, expectations of how the research is carried out and its quality and purpose, and the audience at whom the research is targeted. Practical issues such as time limits, financial resources, available support, style of presentation of the research and word-counts differ, and available guidance by experienced researchers is often dependent on the status that the research is given. Although there are overlaps in what it means to be a researcher in different settings, skills and techniques appropriate to different topics of investigation and different approaches to research can be different.

Above all, though, it can be said that whatever the setting, people become researchers because they have a desire, short-lived or lifelong, to find out more about the world. The advantages it may bring to career development are many but many people who enjoy being a researcher say they do it because of the personal satisfaction it brings them.

In the next sections, we take a brief look at being a researcher in three common settings: academic, practice-based and community settings. Later chapters in the book discuss each of these in more detail.

Being a researcher in an academic setting

Academic settings are most commonly universities, but many clinical training institutions now require academic research to be conducted by trainees as part of professional doctorates and other programmes. Being a researcher in an academic setting usually means you are allocated or you choose a supervisor who guides you and the research until qualification. Remaining in the academic setting often means you combine research with teaching and other tasks associated with working in an academic environment. Researchers in academic settings carry out research alone or as part of a team and are usually expected to attract research funding.

Academic settings also include research fellowships and independent research. Research fellows are supported by the university and are responsible for not only carrying out the research but also for developing research objectives, projects and proposals, and evaluating their outcomes. Research fellows often help (or lead) the development of research funding applications. They have to be able to work both independently and as part of a team, often taking on research team leadership roles such as principal or co-principal investigator. Teaching and supervising students can form part of the role, as can desk research into potential funding sources and literature reviews. Research fellows also write research reports and articles and present their research at conferences.

Working as an independent researcher is more akin to running a business; working on other people's projects means seeking them out and marketing yourself. Independent researchers attract sufficient research funding to pay for their time and other research resources. Their pay is dependent on the financial resources of the project, and there may be periods of working on several different projects or none at all. Benefits in the form of paid leave or sick pay are rare, and independent researchers may not be around other colleagues much of the time. It gives a certain freedom to what they research but can also constrain pursuing personal research interests in order to maintain an income. Many independent researchers relish the excitement and flexibility of working this way because it enables them to cover a range of topics with a number of different people and in a variety of settings.

Carrying out research is part of the university lecturer's role too. The amount of time that is expected to be spent on this varies with the employing university because time must also be given to teaching and administrative tasks, making time-management an essential skill for university lecturers. There are advantages to being a researcher who is also a university lecturer, such as having access to networks and colleagues with which to form teams and generate research ideas. On the other hand, it can be frustrating to have to stop/start research because of other commitments, and there can be pressure to write publications and obtain funding in order to gain promotion. Some lecturers prefer teaching to research but all are expected to keep abreast of research developments in their fields in order to ensure students are benefitting from cutting-edge knowledge.

It is clear that a wide range of personal and professional characteristics are needed to be a successful researcher in academic settings. In addition to the above, it is necessary to be resilient to being unsuccessful in obtaining funding, perseverant with writing and revising papers for publication in high quality journals, flexible in managing more than one research project simultaneously, supportive and nurturing of less experienced researchers and firm in taking projects, and those involved in them, forward to completion.

Depending on the field of research, the working environment may vary from being in an office to being in a laboratory. The environment may be important: having to share an office whilst trying to write can be a challenge to some whilst others thrive on the interactive environment. Finding participants and administering questionnaires, tests and interviews to participants may be daunting for some whilst others prefer human interaction to solitary writing up. Being able to anticipate and plan for research tasks is an essential part of being a successful researcher, as is adjusting timings and plans to fit changes in proposals or direction of research. Balancing all of this with non-research-related work, not to mention personal life, can be challenging and risks, sometimes, being overwhelming.

This, though, is true of many roles and requires characteristics common to many areas of life. We all know of times when we have planned family events, trips away or time to ourselves, only to be thwarted by unexpected events. Being a researcher requires an awareness of the need to be flexible in order to work successfully with colleagues to pursue a common passion and to share the enjoyment of completing a study.

Being a researcher in a professional practice setting

Being a researcher in a professional practice setting means being someone who is employed in a professional capacity but does research as part of their role (Fox, Martin & Green, 2007). In this book, I refer to these researchers as practice-based researchers. Research in professional practice settings ensures accountability, informs evidence-based practice and offers methods of evaluation. The research may be an identified aspect of the practitioner researcher's role, expected either to be carried out alongside other practice-based responsibilities or may be something that the practitioner researcher does on an ad hoc basis. The areas of research may be directly related to professional curiosity, such as why people behave in certain ways, or may be used as a form of assessing and evaluating a service or as a way of incorporating service users' views into development and practice.

One of the challenges to being a practice-based researcher can be that their research is seen as less important than that of academic researchers (Bager-Charleson, 2019). Whilst academic researchers' research is often recognized in the form of promotion and publication, that of the practice-based researcher does not always reach a wider audience (although new practitioner-research based journals are being launched all the time, including the *Journal of Practitioner Research, Practitioner Research in Higher Education* and *European Journal of Qualitative Research in Psychotherapy*).

It can be hard for practitioners to find time for research, and although one might think that they have easier access to potential participants, it is often not ethical to recruit from one's own professional setting. Some practice-based researchers collaborate with academic researchers in order to give the research more space and time, but this carries the inherent constraint of having to conform to expectations of academic research, which may not always be appropriate to practitioner research, in terms of time available or where to publish, for example. The unique perspective brought by the practitioner to the research can be lost when combined with that of academics or other stakeholders in the research unless special care is taken to ensure it is of equal or greater status. Methodologically pluralistic approaches to research can be useful here because they proactively seek to foreground what is of greatest importance to the research question, which in itself can be devised to emphasize the practitioner's area of interest (Frost, 2021).

Becoming a practice-based researcher often develops through training for professional qualifications. Many professional training programmes now incorporate research learning and practice, with an increasing number developing Professional Doctoral levels of study, so more and more practitioners are starting professional life with a background in research. Professional doctorates can be challenging in their requirement in that the research is of doctoral standard yet with a slew of additional requirements to gain practice and supervision alongside theoretical study. The doctorate is usually shorter in wordcount than academic doctorates but this by no means makes it an easier task. The practitioner-researcher has to be someone who can learn to develop feasible yet worthwhile studies and present them concisely yet meaningfully and in depth. The critical thinking and problem-solving aptitude has to be developed towards research and writing as well as towards practice and service development.

Professional doctorates incorporate a research study, developed and carried out by the trainee and assessed to be of doctoral standard. The wordcount is shorter than that of an academic doctorate (typically 40,000–45,000 words), so careful decisions must be made about the amount of data to be collected and analyzed and what to include in the final thesis. The work to be done is no less, however, and has to be included alongside other requirements such as gaining hours of experience and supervision.

Practice-based researchers may also choose to pursue a Doctorate by Public Works. This is most often achieved by experienced practitioners who have produced research reports, published articles, computer programs and other publications during their careers. They systematically analyze these to produce a context statement that critiques the value of the public works to the field. The doctorate is usually supervised by an Internal Supervisor in the institution and an External Supervisor working in the field.

Once the doctorate has been successfully completed and the trainees enter the profession, there is an expectation to inform their practice with research – their own and that of others – so a solid grounding in research is an asset and, increasingly, a requirement of becoming a practitioner.

Characteristics of a successful practitioner researcher, therefore, include not only a passion to conduct research but a commitment to developing it to stay relevant to the area of practice. Working with others is a useful and, often, practically efficient way of ensuring that the research gets done and that you remain dedicated to ensuring it is written for and presented to audiences who will understand it and its relevance to them.

Being a researcher in a community setting

Community-based research has at its core partnerships between researchers and community members. It is collaborative and inclusive and seeks to bring about change through the research. Community-based researchers, therefore, have to be open to insights, views and opinions and to finding ways to include them meaningfully in the research. They also have to be motivated to care about not only the community but also about the topic that they are researching and be prepared to understand different agendas for the carrying out of the research. Training to become a community-based researcher is usually provided by organizations with interests in the community who will work with the researchers and community members to develop the research and its focus.

There are overlaps with the roles and characteristics of researchers in other settings but the approach and passion with which community-based researchers perform their work may originate in different experiences. Often, the interest in being a community-based researcher comes from being a part of the community. There may be issues that perturb the researcher or a recognition that the members of the community are overlooked or oppressed. To ensure that the research is carried out rigorously and in a way that will be persuasive to its wider audience, the community-based researcher may have to be prepared to not only justify the research's value to the community but also to hear the community members' arguments for and against it. The risk of feeling a sense of letting down the community of which they are a member or have formed a partnership with is a real one that can be challenging and hard to accept. If the researcher is a member of the community, the outcomes and consequences of the research and its acceptance more widely may be more value-laden than if they were an outsider to the community. If they continue to live with or belong to the community following the research, they may be in a position to feel a sense of achievement in bringing about change for their community or in a position of having to accept that the collaborative work, which all have engaged in, has not brought about the desired impact.

Conversely, if the community-based researcher is not a member of the community but working with it to develop and carry out the research, they may have to develop trust with its members first, alongside an acceptance that there will be implicit knowledge held by community members that they will not have access to. They may be treated with suspicion or dismissed as being yet another 'do-gooder' seeking to enhance their own credentials. Work must be done to form partnerships, listen to concerns and develop trust. In the community setting, this goes beyond the requirements of research carried out in academic or professional settings, where the relationships formed with gate-keepers and participants may be on a more formal and temporary basis with a clear identification of the role of the researcher.

Often, the community-based researcher is a socially and politically motivated person, who is willing and able to develop relationships with community members known to them in another capacity or not known to them at all. The formal training in research that the community-based researcher may have received often needs to be supplemented with trust-building and rapport skills to maximize participation. Clear communication with community members is key to community-based researchers' research.

Each of these settings for researchers is discussed and explored in depth in Chapters 4, 5 and 6.

Further reading

Eley, A., Wellington, J., Pitts, S., & Biggs, C. (2012). *Becoming a successful early career researcher*. Routledge.

This book recognizes the importance of early career researchers (ECRs) to institutions and to national and international funding bodies, by describing and discussing common features of an ECR's job. It highlights the need to establish a professional identity in order to develop into an independent researcher and provides practical advice to help ECRs kick-start a successful academic career.

Etherington, K. (2004). *Becoming a reflexive researcher: Using our selves in research*. Jessica Kingsley Publishers.

This book shows the reader how reflexive research works in practice and links it with underpinning philosophies. The author includes her own journey as a researcher alongside others and suggests that recognizing the role of self in research can open up opportunities for creative and personal transformations.

Frost, N. (2016). *Practising research: Why you're always part of the research process even when you think you're not*. Palgrave Macmillan.

This book goes beyond the traditional boundaries of quantitative and qualitative research to provide an accessible guide to thinking about the role of the researcher – who they are, what they do and how they shape their own practice.

2 Being a Researcher

Introduction

In Chapter 1 we had a look at some researcher characteristics and qualities that enhance research quality and enjoyment. In this chapter, we will discuss why and how these are important by considering what they mean in research practice in different research environments. Research practices vary according to the research context although all researchers share a common goal of generating new knowledge. Being a researcher, therefore, will be influenced by what is expected of you in the context in which you are carrying out research. However, the requirements to be a researcher differ according to the setting in which the researchers are working. Academic researchers, for example, are usually required to show a track record of research and research publications, whilst practitioner researchers are expected to carry out both professional practice and research interests. Community researchers often combine being a researcher with other roles in the community. In the following sections, we will examine the requirements of being a researcher in each of these settings and consider what this means in practice.

Academic researchers

As we have seen, academia cultivates researchers from an early stage in their academic studies. Undergraduate researchers are required to conduct an empirical or theoretical piece of research for assessment, which contributes to the class of degree awarded. At Master's level, the requirements lean more heavily towards research capabilities and the research proportion of the degree is greater. Doctoral level researchers are usually expected to work largely independently to conduct and write up a research study with support and guidance from one or more supervisors. After this, academic researchers may go on to do research exclusively or in conjunction with other responsibilities.

What does this mean in practice?

All research benefits from researchers having an interest in their research topic. Research studies, whether small- or large-scale, take time and commitment. There needs to be dedication to completing the various stages of the study, to be willing to tackle tricky decisions, to amending ethics applications, to going and learning more about research methods or reconsidering recruitment. All this in addition to, of course, gathering and analyzing the data, theorizing it and then writing the study up. With little interest in the topic you are studying, much of this will seem onerous and something that is taking you away from other, more enjoyable aspects of your studies.

In practice then, many student researchers choose to study a topic that they have had personal experience of or that they have observed in their personal or professional lives. Some new researchers know that they want to pursue a particular topic in depth and see the undergraduate research as the start of an increasingly complex and larger-scale project that they will take on to Master's and then Doctoral level. Others simply have a curiosity that they want to assuage by understanding more and bringing new insight to a topic. Once qualified as a researcher, there can be, initially, less choice in what you research – you may join a research team with the focus of the research already decided or have difficulty in attracting funding for the topic that you want to study. Of course, as experience as a researcher develops, the choices can increase again as you build a team around your interests and become successful at attracting funding.

In all stages of academic research, researchers often have to balance the research with other commitments – completing assignments as a student, teaching and supporting other researchers. A common pitfall is to think that having more time from start to completion of the study means that other assignments can take priority. Any good student will tell you this can be an expensive mistake to make. The academic year is short and moves fast, and there has been many an anxious student putting in late nights and early starts to complete their research. It is also important to factor in time for aspects of the research that are not solely in your control such as applying for ethical approval.

Research cannot proceed without ethical approval, and it is important to remember that it may be necessary to gain approval not only from the university but from other stakeholders as well (such as those in medical or educational institutions). Although many ethics review boards recognize the value of a speedy turnaround of ethics proposals, time taken will be dependent on the number of applications and the number of reviewers available. For students, there are crunch times near the start of the year when many applications are likely to be submitted. It never hurts to try and get ahead of this rush by applying as early as possible, and allowing time for amendments and for the resubmission to be requested. Sometimes non-university institutions take longer to review and approve ethics applications, because review board meetings are held at specified times (perhaps once a month).

In practice, participant recruitment needs to be carefully considered before deciding on the focus of your research study. You may need a large number of participants or, conversely, participants from a small and specialized pool. Although you may be passionate about your research topic, it may not always be easy to find volunteers willing to contribute their time or to share personal experiences with researchers. It is useful to think about the times and financial pressures that being a research participant can entail. What about childcare? The cost of transport to the site of data collection? Taking time out from work and other commitments? Although some funded research enables researchers to offer incentives in the form of travel expenses or childcare, this is often not the case for student researchers. In addition to thinking about your skills and capabilities and how they can be utilized as a researcher, it is important to consider too the time and availability for research that potential participants will have.

On paper, deciding to ask people about their personal experience or to complete a survey can seem straightforward. In practice, it can be affected by your own feelings about approaching and talking with people. If you are a shy person, you may want to think about different ways of gathering data – observation or using online for a survey, for example. If you are someone who likes to develop a rapport with people, then interviewing can be an enjoyable and productive form of data collection. Think about your technology skills too – are you comfortable using computer programs and tests to gather data? Are you familiar with online voice recording of interviews? How will you securely store data online?

Finally, think too about your supervisory relationship. Sometimes, researchers can choose who will support and guide them through the research, but this is not always the case. As a student, you may be assigned a supervisor, and later, you may join a research team with team leaders in place to guide the research. It is to your benefit and to that of the research to develop strong working relationships with supervisors. Use them and their time wisely. This means sticking to agreed deadlines for submitting work and updates to them and preparing well for meetings with any concerns and particular queries. It is important to be able to take criticism constructively and see it as a way to develop your skills as a researcher and maintain the quality of the research. Falling out with a supervisor or team leader can have detrimental effects, ranging from personal disagreements to losing their support. Supervisors and team leaders are experts in research and have its success and quality as their goal. Whilst it can sometimes feel like they are judging you rather than your research, this is rarely the case and it is worth investing time and effort in developing relationships that serve you and the research well. All researchers will be pleased at the conclusion of a strong, high quality study.

Practice-based researchers

Practice-based researchers will have educational and professional qualifications. Their research skills have been proven (or are being honed if they are in a doctoral programme), and the expectation is that when applying for jobs as researcher practitioners, they have an interest in research as well as practice. Job requirements often include the ability to assist with and co-ordinate clinical trials and research studies as well as to deliver high-quality practice. On the one hand, these roles offer great opportunities to investigate the 'burning questions' that arise in practice, to evaluate the practice and to work with multidisciplinary colleagues to pursue wider topics. On the other hand, attention to both aspects of the role can be demanding and, at times, frustrating. Practice-based researchers may have a preference for practice delivery over research, or vice versa, and/or may feel that the research role is regarded as secondary to the practice. Practitioner researchers are often motivated by a desire to address a gap or assumption that they observe in the way their practice is delivered and to improve evidence-based practice in their field more widely. Sometimes, practice-based researchers have to argue the case for developing a study in an area that they are interested in, showing that it is relevant to their work and the practice generally. Other times, they may have to conduct research for the benefit of the institution rather than for their own interest. There are particular challenges for practitioner researchers, ranging from ethical considerations of researching colleagues or clients, to risking reaching results that institutions may not be happy to hear.

What does this mean in practice?

Finding time to carry out both aspects of the role, to be researcher and practitioner, can be challenging, particularly when clinical and research supervision, team meetings, administration and other tasks associated with each are factored in. In practice, this can mean creating and keeping boundaries that allow you to move between tasks required of each role. This can be harder to do when you have a particularly demanding period in your practice – a lot of referrals, a difficult cohort of pupils or emotionally challenging client situations. Similarly, it can be hard to have to take a break from research when it is stuck because it requires decision-making, or is facing recruitment challenges or analysis difficulties, for example. Conversely, if the research is progressing well, you may feel a pull to keep going and not lose the momentum – a choice you may not have if your professional practice calls. A more positive way of looking at the researcher-practitioner role, however, is that it provides variety and support for each role – your practice can inform your research and vice versa.

Although all research must be ethical, and receive approval to proceed from the relevant review board, practice-based researchers often have a number of other steps that they have to take before being authorized. Researchers working in professional practice sometimes have to consider gaining permission for research from a range of other people, including their line manager, senior management, gatekeepers enabling access to participants, and research and development committees, all in addition to the research ethics committee (Fox, Martin & Green, 2007). This is particularly the case for practice-based researchers working in national state bodies, such as the National Health Service in the UK. It is always worth checking with your professional body from the outset of your research to ensure compliance with the relevant bureaucratic and ethical requirements.

Conflicts of interest are a key consideration when conducting research as a practice-based researcher. These may be when including service users in the research and/or when conducting research on behalf of service providers. It is important to include service users when investigating their experiences or evaluating the service they receive. However, researchers must think carefully about how a request to take part in research may be perceived by people they provide a service to. Researchers need to ensure that they make clear how decisions about whether to take part or not may influence participant access to or provision of service to them.

In practice, whilst practice-based researchers are likely to have all the best intentions in assuring potential participants of their freedom of choice to take part or not, the perceived power dynamics between a service provider and a service user can mean that potential participants may nonetheless feel wary of saying no (or indeed saying yes). This can be minimized with clear explanations and offers to discuss, more fully, the implications of taking part in the research. It can be useful to suggest other people whom potential participants can talk to, such as others who have taken part in previous research or other service providers.

Sometimes, gatekeepers are utilized for recruitment. These are people who can enable access to participants and provide a neutral point of contact. They do not necessarily have to be able to identify who has been approached and instead can act for a community of potential participants for whom they can liaise with the researcher. It is, of course, essential to fully brief the gatekeeper not only of the research but also of their role as an access point in it, to ensure bias and miscommunication are reduced.

Another source of potential conflict for practice-based researchers is that the outcomes of the research may be received differently by different stakeholders in it. Think, for example, about an evaluation of a service which concludes that it is not up to standard. The results may be disputed and there may be requests that the study is not publicized. The research may lead to mistrust of the researcher or of the other stakeholders. These situations can be difficult. The researcher will have a responsibility to report the results, but the funders or management of the service may restrict how the study findings are disseminated. To go against this could potentially compromise your position in the organization. It is important to draw up a contract with funders and management before the research begins so that all involved are clear of the extent of their control over what happens to the research on its completion.

Community-based researchers

Community-based researchers are people who conduct research within the community that they are a part of. Often working as part of a team, they conduct research alongside members of government organizations, academic, clinical and education institutions, voluntary and charity sector bodies, social action groups, lay researchers and peer researchers. A key aim is to bridge the gap between communities and researchers by developing partnerships that carry out research with rather than on community members. The particular skills that community researchers bring to the research are in enhancing the possibility of more inclusive research by reaching people and communities who may not otherwise be included in research, having networks and access that those outside the community will not have and being in positions more likely to gain the trust of potential participants.

Community-based researchers are often motivated to conduct research because they want to address social injustices in their groups and bring about change. They can enhance understanding of the lived experiences of community members and illuminate the wider context of the lives and issues affecting them. This, in turn, can increase the likelihood of impactful change being brought about and appropriate policies being developed and taken up by community members.

Community researchers are often involved in all aspects of the research, from its design, data collection and analysis, and have opportunities to gain new skills and knowledge by working collaboratively with all others involved. Research can focus on a range of issues, from public health service access inequalities to oppression within or exclusion from educational support. Communities can be differentiated by a wide variety of factors, including geographic, ethnic, urban/rural, gender, sexuality, age-based, and so on. Above all, community-based researchers are in a privileged position to alert 'outsiders' to previously unseen or unheard community issues, advise stakeholders in the research on appropriate ways for it to be conducted and influence how and to whom the research outcomes are reported and implemented.

What does this mean in practice?

The community-based researcher is often driven by a desire to address inequality and bring about change in their community. This can be because of something that they have personally experienced or witnessed or something they have been made aware of because of their membership of the community. In practice, they will have an identity and place within the community that they wish to capitalize on to benefit the community by spearheading and guiding useful research. Although they are in a strong position to better listen to and understand what other community members are telling them, and to support them in taking part in the research, the role can also bring challenges.

One of these is the very position of being a community 'insider'. Being an 'insider' can mean many things. A community researcher may have a shared experience of ethnicity but have different experiences of health issues, for example. They may be motivated to bring about change in an aspect of the community that not all will want, such as changes in alcohol or tobacco consumption. Insider access can mean recruiting friends and family to take part in the research, but asking and learning about personal or sensitive aspects of their lives that they may not otherwise have known can leave them feeling uncomfortable. There may be challenges to their own cultural beliefs or new insights about people that they thought they knew. Insider researchers may bring assumptions about others in the community that risk being imposed on the data collection and analysis or biases that influence the ways in which data is interpreted or reported. In practice, being an insider on one dimension does not mean community researchers can expect insider knowledge to be taken for granted. It may also be that participants who know the researcher may feel reluctant or uncomfortable to share sensitive confidential information with them for fear of what may be done with it.

All of these considerations of being an insider researcher can affect relationships and status within a community, and this may have influence after the research has ended. Community researchers continue to be members of the community and they will have to make decisions about how to process and manage the information that they have been privy to. They may feel the need to follow up with the participants to ensure their well-being or find ways to help, yet this is action that they are in a position to take only because they have been community researchers. Issues of confidentiality and anonymity must be maintained in community-based research just as it must in academic and practice-based research, but ensuring this is the case can be hard, particularly if the community is a small one. Furthermore, community-based researchers may have relationships that precede the research, and it can be challenging to stand back from or ignore distress that has been communicated during it. A clear ethical awareness and a reflexive approach to understanding the community researcher role and challenges to it can help with this (see more below).

It can happen in practice that the research promoted by and carried out by the community-based researcher to bring about change does not do so. Or that, if it does, it takes time to occur. This can leave the researcher still in the community at the end of the research and experiencing the same issues that prompted them to develop the research. This can feel disempowering and frustrating. Depending on the extent of the community-based researcher's involvement in the different stages of the research, they may or may not be part of its conclusion and writing up. Having collected and been part of the analysis of the data in a collaborative effort with academic and other researchers, community-based researchers may be left alone as the research is brought to its conclusion. Not only does this remove support structures that may have been there to provide platforms for discussion of difficult decisions or emotionally challenging aspects of the research, but it also leaves the researcher in a position of not knowing what is happening with the research nor the timescale in which it is happening. This can be difficult and can also engender a mistrust from the participants that they had encouraged to take part. Finding ways and establishing agreements about how to maintain the communication from the outset of the study to its conclusion can help with this, as can holding regular update meetings with the community and/or keeping them informed via social media.

Reflective question

What are the overlaps and differences you have noticed as you carry out research with researchers from settings different to your own?

Now that we have considered some of the aspects and challenges of being a researcher in academic, practice-based and community-based environments, the chapter will move on to discuss skills and qualities useful to researchers in all environments.

Emotional impact

It is still uncommon to find reference in existing research to emotions evoked in the researcher, but it is an area in which there is increasing recognition of the impact it can have, both on the researcher and the research (see e.g. Moncur, 2013).

There are many reasons for considering the emotional impact of doing research. Whether you are an academic, practice-based or community-based researcher, it is likely that at some point in the research process, you will experience an emotional response. This may come from researching a sensitive topic, feeling isolated or feeling overwhelmed at points. You may experience unexpected emotions or emotions provoked by resonance between participant experiences and your own, and these may add to the emotions you were anticipating.

How you manage effects on your feelings and emotions will affect your enjoyment of being a researcher. Being prepared will always help.

One way of being prepared is to ensure you have appropriate support in place. Having a supervisor or team leader is important to the research, and sometimes, that person can also be a source of support for the more personal aspects of it. If the relationship between you is purely task-based, however, you may want to consider finding a mentor and/or fellow researchers and friends who are willing to hear about some of the emotional challenges that can arise. In addition, having self-awareness of what you are experiencing and finding ways to understand why you may be feeling these emotions will help to identify triggers and strategies for managing their impact. Journalling, mindfulness and creative activities such as drawing can all help to identify and make sense of emotions and their impact.

Ethical requirements often emphasize preparing for the emotional impact of research on participants more than on researchers (Bondi, 2014; Frost, 2016). Researchers are expected to ensure that they work to minimize distress to participants, and ethics applications usually have to show what steps you will take to do this during and after the study duration. These include enabling participants to withdraw from the research or take a break from it if they wish, making clear how the study will be disseminated and safeguarding anonymity and confidentiality. Whilst these are important, caring for yourself is just as crucial.

Researcher self-care includes awareness of the physical environment but also awareness of the potential personal emotional impact of the research. Self-reflection and insight into aspects of the research that may provoke feelings in you will help you to know when to take action to address them. It is important to remember that feelings can be positive as well as negative, so it is useful to think through why you are interested in this topic to research, what you want to gain from it, what you expect to find in it and how open to other insights you may be. Consider, for example, that you are thinking about researching sibling relationships. Why might this be? You can ask yourself about your personal experience – what type of relationships do you have with your siblings? Are you someone who wished to have a sibling? Someone wondering about having a sibling for existing children? Perhaps you work with family relationships? Or are interested in effects of age order of siblings? Then it is useful to extend this thinking to why you are considering formalizing your interest through research – is there a particular aspect of sibling relationships that stands out for you? Are you hoping the study will give you answers to issues within your own sibling relationship(s)? Will the need to delve into existing research in order to prepare for your own give you new insight into your own sibling relationships?

Thinking through these and other questions can help you identify what you are hoping for from your study and your participants. You will be better prepared for feelings of excitement or disappointment and be better able to consider what it is about your own experience that is being provoked in the research. You will be more able to recognize insights that contradict your expectations and the ones that may annoy you or cause resentment. In turn, your passion for and motivation to study the topic will be better supported.

Such questioning of yourself and your expectations should be regularly reviewed throughout the research. You can develop a research journal for questions and explanations of personal reactions during the research to do this. You may benefit from opportunities to discuss with trusted others the feelings that the research is provoking in you. Supervisors will be interested to hear about sticking points in the research and to try and understand with you how these may have arisen from emotional feelings about them. Talking is a useful strategy in itself and can often free you up to move forward in the research. Similarly, writing about your feelings and emotions can be helpful in working through them. It is useful to know if you need to step back and take a break from the research if emotions are too overwhelming or threatening to you or the research.

The following quote is from a researcher journal kept throughout a study of posts written by adolescents in Ireland on an online mental health support forum (Dempsey et al., 2019). It shows researcher emotions triggered during the data analysis that ranged from distress to powerlessness to concern, and how the researcher managed them.

Researcher reflection (from an experienced researcher working with online data for the first time)

I found the sheer volume of data describing despair and confusion to have a strong impact on me. At times it felt close to overwhelming and I took breaks in order to try and read it with a view to analysis, rather than succumbing to feelings of powerlessness in assisting the posters in their plight. I felt concerned for the young people writing the posts and wondered whether they managed to find the help they were seeking. I also wondered whether they felt sufficiently empowered by writing the posts to accept and act on the advice provided.

Managing time

Managing your time as a researcher means organizing your commitments to ensure that research takes its place in the list so that submission and funding deadlines are met. Many researchers are carrying out research alongside other commitments, and all are doing it in the context of living their life. The nature of being a researcher means that you have to plan for quieter times in the research process (waiting for ethical approval, for example) as well as for the busier ones (often the writing-up and publication stage). Sometimes, it can feel that for all your careful planning, the process is out of your control and joins other aspects of your life demanding attention at the same time.

Time management needs to be considered from the research planning stage right through to writing up the research. Drawing up a realistic timetable can better allow for this. How will you fit the research around your teaching or practice, for example? What about the need to travel distances to gather data in the field? How will you find the hours to be alone to write it up without other distractions? Answering these questions may mean that you have to call on the support of colleagues, friends and family, perhaps needing cover for your teaching or help with childcare, for example. Whilst it may feel difficult to ask for this support, most people understand the need for you to pursue your research interests and will be pleased to be part of it. It also means they can call on you to return the favour to help with their own projects!

No amount of planning will allow for the unexpected events that occur in the research process but at least with a realistic timetable and some built-in flexibility, there is a good chance that you will be able to complete on time and gain the satisfaction of generating new knowledge that will add to your field of interest.

Acquiring relevant skills

Knowing that you *want* to do research, and the area you want to do it in, is the first step in being a successful researcher. Knowing *how* to do it is a continuing learning curve and can lead to new challenges. You are likely to need new knowledge or skills to progress the research, and these can range from having to learn new research methods to adapting your writing skills for different audiences.

Most researchers are offered training in the how to of research – what steps are needed, which methods are available, the requirements of writing up and so forth. This, with some practice, can help in getting going with planning and getting the research off the ground. However, the complex nature of real-life research means that a range of additional skills are required and, further, that these cannot always be taught.

Some of the skills that can be learnt include critical reading of existing research in order to identify a gap in the existing research, understanding ontological and epistemological underpinnings and learning analytical data techniques. Knowing what to do does not always equate to knowing how to do it, particularly for researchers early in their career, but is certainly an important step. Skills that are less easy to be learnt through teaching and more through experience include how to approach potential participants, how to elicit meaningful data and how to recognize bias. These rely on researcher characteristics, communication skills and confidence in knowing how you have acquired beliefs and values to supplement what you have been taught.

It is useful, therefore, to consider which skills you have, which are the most important for you to acquire and how to apply what you have already learnt from your experience as a researcher.

There are plenty of training courses available to help researchers learn new methods and approaches to research. Organizations exist, such as the National Centre for Research Methods (NCRM) in the UK that runs an international Research Methods Festival every year (ncrm.ac.uk), which covers all aspects of research. Many clinical and academic institutions also offer specialized courses. Most cost money, so it is important to think about the value of attending, and it can be helpful to find out whether your institution may help with costs. Some researchers in need of a particular skill, perhaps in a certain method, for example, get together to make a case to their institution to bring specialized training to them. Additionally, conferences often offer pre-conference workshops delivered by experts in specialized aspects of research.

It may be, however, that sometimes researchers have to recognize that they do not have a particular skill needed for the research – knowing how to approach a particular community, for example – or not having expertise in a particular research approach. It may be pertinent to question closely whether this means that you are not the right researcher for this study, whether you need to bring someone else into it (finding someone who has easier access to a community than you do, for example) or whether it is a skill that you can acquire. This can save wasting time on a research study that cannot recruit participants or does not address the central question or hypothesis appropriately. If the approach and method(s) that you are skilled in will answer a different question, it may be useful to think about refocusing the study. This decision should not be seen as a failure but as a responsible and useful approach to research – there is no value in poor quality research, but each piece of good research adds to knowledge. The key point is to decide what skills you have and which you need as a researcher for this research study.

It is worth highlighting here that skills such as leadership are also important in research. As you become more experienced and immersed in research, the likelihood of you leading a team increases. This is a skill that can make or break good research. Ways of building this skill can include working with Research Assistants, writing collaboratively with others, supervising research students, reviewing manuscripts for journals and examining doctoral research. Such roles will show you how working with and guiding others can enhance the research quality and progress until the day comes when you feel confident enough to propose or take up an official Principal Investigator or Team Leader role.

Supervisory and support relationships

The importance of having people you can turn to for advice and guidance as a researcher cannot be overemphasized. A recent study (Bager-Charleson & McBeath, 2021) found that 71 per cent of trainee counsellors and psychotherapists rated good supervision as key to good research.

It is common for student and trainee researchers to be allocated one, sometimes two, supervisors, whilst postdoctoral researchers more commonly have experienced research team leaders to discuss and guide the research with them. The role of supervisory support is a broad one, ranging from instruction (for novice researchers) through to enabling opportunities for discussion and problem solution. Pastoral care has a variable role but empathy, supportiveness, flexibility and sensitivity to researchers' needs are widely recognized as important to the success of the relationship as well as to the research (Bager-Charleson & McBeath, 2021).

Supervisors' key responsibilities include ensuring that research and administrative functions are carried out appropriately and in a timely manner. For many researchers starting out in research, it is important to recognize the need for this guidance as much as recognizing the need for conducting high quality research, because it prepares them for future funded research. In practice, supervisors draw on their topic knowledge and methodological expertise to guide researchers through the process, whilst also using their knowledge of the university, clinical and other institutional expectations to ensure the appropriate paperwork is completed. The relationship with them is, therefore, a very important one, and to develop it well, it is critical to recognize that supervisees have responsibilities too. The box below outlines some of the key ones.

Supervisee responsibilities

- Seeking help in a timely manner when needed in order to ensure personal and professional development
- Maintaining regular contact with supervisors and responding to communications in a timely manner
- Preparing for supervisor meetings
- Meeting deadlines set with the supervisor
- Familiarizing yourself with the institutional administrative expectations and working with the supervisor to meet these
- Keeping supervisors informed of issues likely to affect the progress of the research.

Written out like this, the list may make the contribution of the supervised researcher to the supervisory relationship seem straightforward and easy, but given that the breakdown of these relationships is not uncommon, it isn't a surprise that the challenges can be more complex. Supervisor–researcher breakdown is often cited as the reason for student researchers to discontinue their studies (Fossli & Michaelsen, 2017). Student researchers can feel unsupported by busy supervisors or feel that their supervisors are overly critical of their efforts. This can make it hard to raise personal or research-focused problems with them, leading to the research suffering and the student losing faith in their ability to do the research. Supervisors cite overload and not being kept informed of progress by student researchers as reasons to dislike or discontinue supervision (Parker-Jenkins, 2018). However, there are many success stories of the supervisory relationship, with experienced researchers recalling many years later the value of their doctoral supervisor's support. Like any partnership, effort is needed on both sides to ensure a productive and beneficial relationship, and the key to this is often communication.

Being clear about what is expected of you by your supervisor is an important foundation to a successful relationship. Asking about it and querying any areas of uncertainty or concern from the beginning can be helpful and sets up a rapport that enables further talks when the research poses you with new problems. Having a good rapport with your supervisor also helps when you want to celebrate exciting outcomes or overcoming a challenge.

This approach can also make it easier to talk to supervisors when you are struggling and to develop a plan of action with them. Some supervisory relationships, particularly those that last through years of research, develop into friendships, but at the start of your relationship with a supervisor, you are the student looking for guidance from a respected expert. By the end of it, you should be the expert with a proud supervisor who has supported you to enter the world of research independently. Along the way, you will have navigated challenges and threats to the research and, sometimes, some personal life problems. The extent to which you want your supervisor to know about what is happening in your personal life is up to you and is determined by the relationship you have, but it is always important to let them know if you are struggling to keep on top of the research.

This can be further helped by having others to whom you can turn, with whom you feel more comfortable talking about threats posed by personal issues. Sometimes knowing other researchers can be invaluable in discussing ways to address the threats to the research, but it can also be beneficial to have friends who are not researchers, who know you are doing research and have an interest in hearing about it. Whilst your supervisor holds the reins in sanctioning requests such as taking a break from the research and should, therefore, always be kept informed, friends can help you to disentangle the personal from the academic by finding possible solutions. If that solution appears to be to take a break from your research for a while, that is the time to talk to your supervisor.

Sometimes, researchers, particularly more experienced ones, are located away from institutions where their supervisor is based. This can particularly be the case for community researchers who may be working in communities that are located in other countries. This can make it more difficult to stay in touch or to turn to supervisors for guidance. A way of addressing this is to agree on a timetable of regular meetings to check in and update each other, using technology if necessary and available. Setting up WhatsApp groups with fellow team members can also help with more informal contact and support.

Remember too that although you have one or two supervisors, they may have several supervisees. This can be a little hard to take when you are panicking about arranging a meeting or waiting for feedback on work submitted, but remembering this and maintaining open channels of communication can mean that trust can be established and deadlines agreed upon and met.

Gaining funding

Obtaining funding for research becomes an increasingly important aspect of the researcher's life as they gain experience. Once qualified, researchers often have to consider how to fund their research. This can mean making a case to universities, practice institutions, research centres, and national and international funding bodies that you are the best person to carry out the identified areas of research. Sometimes, experienced researchers have access to funding for particular projects and seek to recruit doctoral students and other researchers to form a team with. Researchers who are part of a team or who choose to create a team will be looking to convince funding bodies that they will use the money wisely and in accordance with the funding body's aims. Furthermore, that they will meet the deadlines and produce research that will be useful. This usually needs to be conveyed in a Funding Grant Application.

Writing grant applications is time-consuming and a skill often best gained through experience and by learning from others. There is a language and a broad format for what funders are looking for, and, of course, the process is competitive, so everyone else applying for the same money will know this. A good piece of advice is to seek out someone familiar with and successful at obtaining money for research and to work with them as your mentor on drafting and re-drafting applications. This may mean working on a project that you feel less passionate about, but the value lies in being able to one day be successful in gaining funding for the study you really want to do. As a researcher, you can bring knowledge of a topic but also of methodology, recruitment or particular communities, so think carefully about what you can offer to a potential mentor before approaching them.

The large-scale funding grants available often require completion of detailed application forms and can take place over two or three rounds of elimination. There may be two or more written applications needed and it is not unusual for shortlisted candidates to be asked for group or individual interviews before a final decision is made. This means that skills beyond writing are needed so that researchers can communicate not only their detailed knowledge and plans for the potential study but also their ability to work with teams, communities and individuals. It is also useful to show that they are connected to networks that will enhance the progress and dissemination of the research.

It is useful then to start early and small when learning to apply for funding. This means keeping an eye out for potential funders for the areas of research you are interested in. Look out for local charities seeking research in the area or for money made available within your institution. Regulatory bodies also offer rounds of funding. Whilst the quality of the research, and your plans for it, must be no less detailed than if you were applying to wider bodies or for more money, the pool of competitors may be smaller and the information needed at the outset less detailed in some areas (for example, charities may be interested in exploring a well-defined issue and less interested in whether you are going to use this method or that than whether you have an understanding of the issue or the people who may be experiencing it). If you are applying within your institution, there may be support available from departments and teams whose role is to look over budgets, and it is always useful, often necessary, to have their authority on the application. Heads of departments often have to sign off on applications, so making sure that they are kept informed and, if appropriate, asking them for their input on it can pay dividends.

Make sure, too, that you are on the various automated lists that are sent out to inform researchers of upcoming grants. Most institutions have them but there are also national international resources, e.g. scientifyRESEARCH.org.

Gaining funding relies as much on timing and a certain degree of luck as on the skill of completing the application. It can be a dispiriting aspect of being a researcher to spend a lot of time on applying for funding with no success, but it is always useful to have the learning opportunity and the feedback, formal and informal, to help towards the next application. For some researchers, it can enhance career promotion, and for others, it is a vital necessity to have an income stream that enables them to continue as a researcher. It is useful to remind yourself why the research is important to you and what you hope it will contribute. Find a support group (online, if necessary) to help keep the motivation going… and celebrate when you are successful before the work of doing the research begins!

Chapter summary

In this chapter, we have considered what it can mean, in principle and in practice, to be a researcher in different settings. We have considered how roles and expectations differ and the skills that researchers need to have or acquire. Along the way, we have thought about some personal impacts as well as practical issues that are necessary to being a researcher. All of these will be recurring themes throughout the book, and we will be returning to them in different contexts from now on. First, though, in Chapter 3 we will consider the role of enjoyment of research for a successful researcher and what difference it makes to enjoy, or not enjoy, what you do.

Further reading

Boynton, P. (2020). *Being well in academia: Ways to feel stronger, safer and more connected*, Routledge.
This book is for those who want to address the need to stay well in academia. It provides practical tips and guidance on ways to develop and provide support to enhance confidence and safety across the diversity of academia.

Curry, D., & Wells, S. J. (2013*). An organic inquiry primer for the novice researcher: A sacred approach to disciplined knowing*. Infinity Publishing.
This book describes the use of a six-principle, three-part process for conducting the transpersonal/feminist methodology of organic inquiry. It is especially well-suited for investigating elusive topics and following questions that take researchers into the unknown.

Ghezzi, C. (2020). *Being a researcher: An informatics perspective*. Springer.
This book is philosophical and, at times, anecdotal, in combining factual information and commonly accepted knowledge on research and its methods, whilst at the same time clearly distinguishing between objective and factual concepts and data, and subjective considerations.

3   Researcher Challenges

Introduction

In Chapters 1 and 2 we have looked at some of the requirements, skills and expectations of being a researcher. In this chapter, we will consider how these take their place in decisions about becoming and staying a researcher. We will look at how challenges of integrating research into other aspects of study, work and professional practice influence enjoyment of it, and what it means to be a researcher who cannot do research all the time. The chapter will end by outlining some of the resources available to support researchers.

Integrating research into academic practice

There can be a perception that working as an academic means that the researcher role is 'easier' because of the requirement to do research and time allocated for this. In addition, academic researchers are often surrounded by colleagues doing the same, which can give a sense of belonging and support. However, it can also be the case that the researcher in you is dominated by teaching and administrative requirements, lessening your enjoyment of doing research.

Combining research with teaching

Most academics are expected to teach, carry out administration *and* do research. Many universities expect between 30–40 per cent of lecturers' time to be spent on research, 30–40 per cent to be spent on teaching and 20–30 per cent to be spent on administrative tasks. Of course, the research element of the role includes all aspects of doing research – developing proposals, applying for ethical approval, carrying out literature reviews, collecting and analyzing data, writing up, submitting manuscripts for peer review and publication and communicating research to communities and conferences. As an academic researcher, you may also be applying for funding, managing budgets, organizing and attending team meetings and maintaining minutes and other audit trails of the research process. To carry out meaningful research, you must also keep abreast of emerging research in your research area by reading and reviewing new publications. You may be asked to appraise funding applications, to review journal manuscripts and to supervise PhD students. All of these tasks can easily take up the majority of the 30–40 per cent of your time.

Teaching requires preparing as well as delivering lectures, setting and marking assignments and exams, supervising student research projects and providing pastoral support. Students and student success are the key focus of most universities, and with the introduction of fees in some countries, there are increasing expectations from students that they have attention from their lecturers and finish their studies with a high class of degree. Whilst the majority of academic lecturers embrace the teaching aspect of their job, there are challenges in combining it with their research.

An obvious challenge is the imposition of a timetable over which you may not have control. The teaching timetable can be heavier at certain times of the year as the modules are rolled out, meaning that you may have less time to devote to research at these times. Additionally, the timetable can mean broken days and weeks with lectures spaced out across them, or conversely, time-tabled back-to-back with no time for research.

It may be that you can talk to the department head (or whoever sets the timetable) to outline your research plans for the academic year and request that your lectures are grouped together over two or three days of the week. If this is enabled, the 'teaching days' can also be the admin days, leaving the rest of the week solely for research. It may also be possible to make clear on your office door and emails which days you are available for student consultations and enquiries. Some universities have an online system for students to book time with their lecturers, meaning that lecturers can block out time for their research.

Another way to make time for research is to factor it into non-term time. This is less desirable to some researchers who prefer to keep the diversity of research and teaching continuous through the term. It also does not always work with the research study itself that may need to be conducted at certain times or take longer than a teaching break. However, it can be that using teaching breaks for writing up and submitting manuscripts for publication works well and, for those that enjoy writing, is an enjoyable way of working to your own timetable.

Finally, an efficient and valuable way to integrate research with teaching is to practise research-based teaching. This will necessitate you keeping up with and selecting key new research in the areas you teach in to inform your lectures and referencing. Encouraging students to question and discuss the research will substantiate the teaching and enable you to further think it through critically. It can quite possibly lead to new avenues for your own research.

Many of these issues are common to practice-based researchers who provide services and products to clients, and we will consider how research can be carried out alongside these requirements in the next section.

Combining research with professional practice

Being a practice-based researcher means combining services and products you provide to clients with doing research. There will be some overlapping challenges with those of being an academic researcher, in that you may have timetables imposed on you if your role is client-facing or demands of the practice may overwhelm your time to research. There are also challenges arising from researching problems of practice or the organization where the practice takes place. These may include research evaluations that risk producing results that the organization or its practitioners do not want to know about or having proposed research areas blocked by those with power in the organization. If the organization does not have a research culture, the capacity within its infrastructure to conduct research may be low or it may not see the benefits of involving practitioners in research. The active practice-based researcher, however, can benefit themselves and the practice through their research by developing, accelerating and extending new practices within and outside the organization, enhancing the likelihood of research findings being implemented and increasing their fulfilment in their work role.

Perhaps, at the forefront of the challenges to being a practice-based researcher is the need to have the support for research from those responsible for managing the organization and its professional practice. Although many organizations offer evidence-based practice, it is not always the case that this evidence comes from within their own organization. Instead, managers and others are content with those working there to follow and sometimes implement research carried out elsewhere. In turn, this can mean that colleagues who do not have the same level of interest in doing research as you do, perceive it as a difficult and unsupported additional task. However, it has been found that a research culture within organizations can improve job satisfaction, lower job-related stress and reduce staff turnover (Pinto, Spector & Rahman, 2019).

To encourage the development of a research culture, it may be possible to speak with colleagues and managers to highlight the value of in-house research. It can be emphasized that having active researchers in an organization not only offers ways of engaging with the community but also can extend to liaisoning with and influencing policymakers, practitioners and researchers from other organizations. Furthermore, if you are a confident or experienced researcher, you can offer to support colleagues in learning about and conducting research so that they feel that they can play a meaningful part. Setting up research discussion groups can enable practitioners to contribute to identifying research problems and developing research questions, have an input on research agendas, inform service priorities identified through research and learn how to interpret results. You can argue, too, that in many organizations, particularly in health and social care, there is an ethical obligation to find ways of promoting social justice and, for some, an advocacy role for individuals, communities and populations who are underserved, disadvantaged or oppressed (Pinto, Spector & Rahman, 2019).

Nonetheless, despite these arguments, there can be challenges in recruiting colleagues to join you in being a researcher. They may feel that they don't have the time or necessary skills or may only be interested in delivering the professional service. Additionally, there may be a reluctance for practitioners to be supported by the organization to carry out research for fear of taking them away from the practice delivery. In these instances, it can be helpful to look outside the organization for other researchers with whom to form partnerships. These may be colleagues in other organizations, universities or community-based researchers. It may also be possible to network with researchers who have a stakeholder interest in particular research areas to work together to develop different services. Benefits of these options include developing pluralistic insights and perspectives on problems identified in your own or others' organizations (Frost, 2021) and having extended reach for the dissemination of your research and its outcomes when it is complete. Professional knowledge can be shared, different participant pools identified and knowledge of funding sources pooled. As described in the sections on team-working as a researcher throughout this book, the success and productivity of working with researchers from different disciplines are enhanced by acknowledging their agendas, availability and knowledge in relation to your own, as well as ensuring communication is regular and in a language you can all understand.

Finally, by joining (rather than forming) a team, it may be that you can have roles and participation in the research that meet your interest and availability as a researcher. It may be that you have quieter times of the year with regards to your service delivery or that you most enjoy (and are more competent in) particular aspects of research, such as data collection or analysis. You may prefer to interact with participants or to work with the data. You may be nervous about leading on writing up, so you can take a proofreader or editor role instead.

Above all, the best way to be a practice-based researcher is to ensure you make time for it and commit to it and that you seek support – from senior members of the organization as well as from colleagues within it and at other organizations. This will enhance the likelihood of the research being completed and disseminated and may well carry other benefits such as networking development and career promotion.

Integrating research with community involvement

Community-based researchers often seem to be portrayed in isolation, as if they exist only as researchers. Of course, like all researchers, they have lives and, sometimes, other jobs to maintain whilst they do the research. For community-based researchers, some challenges can arise from continuing to be an active community member. You may face questioning about why you are doing the research and what the implications of its outcomes might be. You may be faced with hostility if there is contention over the research topic or who has commissioned it. If community members have had bad experiences with researchers in the past, or feel 'over-researched', they may resist taking part despite your involvement. Even if you are a member of the community, you may be regarded differently in your researcher role, perhaps perceived as having greater (and unfair) advantage or status than other community members. This can be challenging on many levels and can be compounded by needing to ask for personal data or personal experiences that community participants may not expect or want to share with you.

Some of these challenges can be addressed by considering how to reduce the perceived power hierarchy that you as a researcher occupy in the community. It is important and valuable to be reflexive in your community research. Consider the different ways in which you might be perceived and how you might address them. This can mean reassuring participants about how the information that they give you will be stored and kept confidential, but also, identify with them to better communicate how they can trust you not to share information. This of course is more than a tick box exercise. Consider how you will develop a research rapport with participants that you may know from the community – perhaps by explaining why you are doing it and what benefits you see as coming from it, how you have been trained and what it would mean to you to carry it out successfully. This may assure potential participants of the high standards you have set for your research and for yourself as a researcher. You may want an open discussion with them about how your existing relationship will continue, allowing them to raise concerns about the impact on the relationship of taking part in the research. Allow potential participants to ask you questions about yourself and your researcher role, and consider in advance what you expect to be asked and how you will respond. Make clear to participants that the research outcomes will be shared with them (perhaps discussed prior to being written up) and who else will have access to the research write-ups. As a reflexive community-based researcher, you should also consider how you will process and manage surprising or distressing information you receive. Think about how it may affect your existing relationship with the participant and, if relevant, with other members of the community.

Another way to enhance your integration of being a community-based researcher with being a community member is to conduct participatory research, in which you involve members of the community as co-researchers. Community Based Participatory Research (CBPR) is the creation of partnerships with members of the community that is meant to benefit from the research. By consulting with community members, knowledge can be shared and developed, allowing questioning of assumptions about the community, its members and the topic under inquiry. Involving community members in the research design – and conduct, if they wish – means the ensuing research is more likely to be better focused and to elicit data of greater meaning and interest to the community. In turn, this can mean that any suggested outcomes from the research may be more effectively taken up by the community.

It can also be beneficial to develop a Research Advisory Board involving community members (Pinto, Spector, Rahman & Gastolomendo, 2015). Community members can be recruited as Research Advisory Board members, alongside academic researchers and other stakeholders, offering guidance and feedback on all aspects of the research. They can also serve as points of contact when decisions have to be made or challenges arise in the research process.

The long-term nature of Research Advisory Board membership can, however, be an onerous request to make of community members. Branney, Strickland, Darby, White and Jain (2017) have developed a short-term approach in which one-day workshops are held with community members and others (perhaps service providers or developers) to elicit information about the topic, to be used to design the research and effective data collection techniques. By emphasizing rapport-building, autonomy and rationality in a time-limited approach, the workshop can minimize some of the challenges of long-term Advisory Board recruitment and operation by reducing long-term commitment to it and helping to minimize perceived power hierarchies between community-based researchers and community members (Pinto, Spector, Rahman & Gastolomendo, 2015).

In the chapter so far, we have considered ways of integrating research into other roles. Whilst it is hoped that these practical tips will help, there remains the question of how to manage being, and what it feels like to be, a researcher who cannot do research all the time. In the next section, we explore this and consider why it matters.

The frustrated researcher

The *Cambridge English Dictionary* defines frustration as 'the feeling of being annoyed or less confident because you cannot achieve what you want' (dictionary.cambridge.org). Frustration can manifest as disappointment and anger, and can lead to recklessness; Giles (2005) found that researchers frustrated by delays in waiting for ethical approval sometimes broke the rules by collecting data before approval had been granted. We can understand why researchers who have many other professional demands on their time see taking short cuts in research as inviting, particularly if the research is on a tight deadline or at a particularly critical stage, but there is never a justification for it. Patience and using time wisely, such as by completing literature reviews or learning more about the methods to be used, can be helpful in reducing frustration with apparent lack of progress in the research.

Feeling less confident in the research, or in your ability to do the research, can also come with frustration. It may be that taking longer breaks from it than expected can lead to feeling less assured about your ability to pick up the research again. You may feel that the skills you had developed have diminished or that you will have lost participants who had previously agreed to take part. Your motivation for the research may be affected so that it does not seem as important as it once was or that its originality has been threatened by other research studies. One danger with these feelings is that the research doesn't get completed or that you start new research projects with equally doomed prospects.

Frustration at not being able to spend the time you want with your research can also lead to breakdown in communication with research colleagues. The importance of communication in teams has been well documented (e.g. Frost, 2016). Successful team membership necessitates an openness with other team members so that all researchers involved are clear about what is happening in the research process and what is expected from them and of others. This is best done with regular contact, in review meetings and online contact. A researcher whose role in the research is held up may feel embarrassed or ashamed at not having carried out their part of the process as agreed and may absent themselves from these communications rather than share their lack of progress. This is damaging to the rest of the team, both in that it risks the researcher's reputation being damaged and stalls the research project.

Having to take breaks in your research can sometimes come as a requirement from the management of your organization. They may not see the value of your research, may see it as interfering in the smooth running of the organization or simply do not want to release the time you would otherwise spend on practice. You may look elsewhere for support, which in turn may provoke the ire of management. In turn, this may mean that access to participants and applications of the research outcomes may be impacted.

It may be that not only management but colleagues too are resistant to research being carried out. They may feel that they are being watched or monitored by the research process or that their participation may take up more of their valuable time. As the researcher, you may start to feel isolated or resented.

Frustration can also grow from a reluctance of clients to take part in research with you. There are clear challenges and ethics to recruiting participants that you have a professional relationship with. They may feel coerced into agreeing for fear of losing your services or may be concerned about confidentiality and anonymity being breached. Being aware of these, and ensuring that potential participants are aware that you have taken responsible, ethical decisions and maintaining clear and open communication with them about breaks in the research and timing of planned returns to it can develop trust and reassurance and provide opportunities for potential participants to raise any concerns they may have.

Reflective question

What frustrates you in your role as a researcher? How does this manifest in your research activity?

Enjoying and not enjoying being a researcher

Becoming a researcher starts for many in early career study and for others in training associated with employment development. For some students and trainees, this course requirement is the least enjoyable, for others, the start of an exciting journey into a research career. As researchers become more experienced, their confidence grows and, with it, the enjoyment of what they are doing. As research careers develop, and individual research studies progress, enjoyment of being a researcher may still waver, but with a curiosity for the world that can be harnessed into doing research, the researcher role is often both enjoyable and fulfilling.

Being a researcher offers opportunities to develop skills in critical thinking, rational decision-making and self-reflection. It can be a challenge for teachers and supervisors of novice researchers to find ways of making research come alive, so they have to think of ways to catch interest, encourage engagement and support progress. To do this, they encourage interaction, highlight the many ways of carrying out research, use examples of their own and other's research and give opportunities for students to practise being a researcher. The trick for teachers and students together seems to be in cultivating inquisitiveness. If this is achieved, researchers will talk of the thrill and the challenge of research, and be excited to get going on it. Nonetheless, there are also researchers who do not enjoy research or who stop enjoying it. They become defeated by the length of time it takes, lose belief in and motivation for it and become frustrated by administrative tasks, tight deadlines and challenges of getting research published. All of this can lead to burnout, poor research quality and abandonment of research.

Burnout is a prolonged response to chronic emotional and interpersonal stressors in a job. It is characterized by the three dimensions of exhaustion, cynicism and reduced professional efficacy (Maslach, Schaufeli & Leiter, 2001). Researchers appear to be at risk of burnout because of high stress levels arising in part from publication pressure (Miller, Taylor & Bedeian, 2011), competitive work environments (Levecque, Anseel, De Beuckelaer, Van der Heyden & Gisle, 2017), high workloads (Barkhuizen, Rothmann & Van De Vijver, 2014) and job insecurity (Guidetti, Converso, Di Fiore & Viotti, 2022). The COVID-19 pandemic added to stress levels and anxiety because research designs had to be modified in light of cancelled or postponed experiments and fieldtrips, and restricted data collection opportunities, increasing the risk of burnout (Gewin, 2021). Some 80 per cent of postdoctoral researchers believed at the start of the pandemic that it was seriously hindering their research (Woolston, 2020). A recent study (Boone, Vander Elst, Vandenbroeck & Godderis, 2022) aimed to identify burnout profiles amongst young researchers in five Flemish universities, and it is summarized below.

Research example summarized from Boone, Vander Elst, Vandenbroeck & Godderis (2022)

Hypotheses:

1a) the first latent profile will show high scores on all burnout dimensions (High Burnout Risk)

1b) the second latent profile will show high scores on cynicism and low scores on emotional exhaustion and professional efficacy (Cynical)

1c) the third latent profile will show high scores on emotional exhaustion and low scores on cynicism and professional efficacy (Overextended)

1d) the fourth latent profile will show high scores on reduced professional efficacy and low scores on emotional exhaustion and cynicism (Ineffective)

1e) the fifth latent profile will show low scores on each of the three burnout dimensions of emotional exhaustion, cynicism and reduced professional efficacy (No Burnout Risk)

Participants: recruited from five Flemish universities (N = 1,465 of which 1,123 met the inclusion/exclusion criteria) using a convenience sampling strategy in which the doctoral schools were asked to distribute a web-based survey to PhD students and postdoctoral researchers. Some 69 per cent of respondents identified as female.

Instruments: the *16-item Maslach Inventory-General Survey* (1981) was used to measure Burnout Risk. A five-point Likert Scale ('never' to 'every day') was used to measure the frequency with which respondents experienced feelings related to emotional exhaustion, cynicism and reduced professional efficacy.
The *Copenhagen Psychosocial Questionnaire* (Burr et al., 2019) was used with a five-point Likert Scale as a predictor to measure job demands, workload and work–life interference.
The *Publication Pressure Questionnaire* (Haven, de Goede, Tijdink & Oort, 2019) and Job Insecurity Scale (Vander Elst et al. 2013) were also used.

Results and implications: supported existing research that the three burnout dimensions manifest differently in individuals. The Ineffective profile (high on reduced professional efficacy only) was not found but a Low Burnout Risk profile that showed low scores on all three dimensions was, suggesting that reduced professional efficacy is not the main problem experienced by young researchers.
The Overextended and Cynical profiles suggested that a subgroup of the sample feels exhausted whilst maintaining low levels of cynicism and reduced professional efficacy and another feels cynical about their work alongside low levels of exhaustion and reduced professional efficacy. In contrast to existing research, this study found three times more people in the Cynical profile and three times fewer in the Overextended profile. This suggests that this participant sample is at greater risk of becoming cynical than exhausted.

This sobering profile of the levels of stress that young researchers report should be noted and addressed from early in their careers. Emphasis on the need for self-care and protection of mental health should be part of researcher training, and the need for researchers to take steps to protect themselves should be highlighted. Universities and other research settings have a responsibility to promote and endorse this to ensure that careers and personal experiences of researchers are protected and supported.

Researchers also report their enjoyment of the role (e.g. Mantai, 2017) and also the ways in which to promote enjoyment of being a researcher (e.g. Heinrich, Hill, Kelder & Picard, 2024).

Experienced researchers describe enjoying research as harnessing a natural enthusiasm for 'finding out' (Bager-Charleson & McBeath, 2020), adventure (Willig, 2021) and developing methods (e.g. Charmaz, 2014). Enjoying research comes from looking at things in new ways, challenging oneself to think outside the obvious and being confident that the knowledge generated is new and credible (Frost, 2016, 2021). Enjoyment is personal, sometimes emotional and increasingly intuitive as researchers become more experienced. The sense of following a hunch by drawing on a repertoire of knowledge and experience, learning from others and reaching an outcome that is interesting and useful can give a sense of achievement. Researchers also describe the enjoyment of being successful in attracting funding for their research, giving it value and status. Many enjoy meetings with participants to gather data from them, opportunities to share and develop their research and the satisfaction of knowing what they are doing is worthwhile and of high quality.

However, the long-term nature of many research studies can be daunting for some researchers, particularly those at the start of the researcher journey. They struggle to see the reward that can come at the end of a research study and weariness sets in. Researchers can start to question whether the study will ever be completed or believe that after all the work, a solution to the problem that they are researching will not be found. Researchers who feel most passionately that their research explores a problem that needs to be investigated are usually the ones with the most determination to see it through, although some become resigned to the research being a job that has to be done. The following reflection illustrates how one researcher navigated ambivalence towards her research.

Researcher reflection (from a postdoctoral researcher working in industry and academia)

I don't know if anyone has ever come to hate their research? PhDs are supposed to be about the burning question that keeps you awake at night and that you are intrinsically motivated by. My PhD question, however, was a bit of a mish-mash between my interests, my supervisors', and an organization whose services, we were evaluating as part of the PhD. Somewhere, along the path, I lost my 'why' with the research, and at times felt the research could be co-opted into what the organization wanted, rather than the constructive critical evaluation we set out to achieve. Did I really hate my research? I think I hated the organizational silences more, and the challenges I encountered in trying to articulate them....but I know that the journey taught me so much, and that with time and distance, I will be able to write my way out of it.

There can also be a fear that the research outcomes will not be applied or that their impact will not be recognized. Areas of research in issues that are not widely known about or are not considered important can cause researchers to question how or whether the outcomes will ever be read or implemented because, although important to them, the topic under investigation is not seen as important to those with power to implement outcomes.

The pressure to publish research and the length of time it can take from writing a paper to its publication can reduce enjoyment of the research. Some researchers prefer the fieldwork and data analysis aspects of research to the task of writing it up. The different skill sets required for writing can feel challenging though, and the time taken can cause concern that advances will have been made before their own work in the area is published. Whilst this concern can motivate researchers to persevere with writing, the link between promotion prospects and research can be overwhelming and reduce enjoyment of the research itself.

Personal characteristics can also influence enjoyment (and lack of enjoyment) of research. Being a researcher can consume many hours in both carrying out tasks and mental activity. It requires commitment and determination, ability to see through problems and curiosity. For some the domination of their time and thoughts can be a double-edged sword: the fulfilment that comes from immersion in a research project can run alongside guilt at the project taking them away from their family and social life. Researchers also speak of the isolation of researching alone as a hindrance to their enjoyment of doing research, sometimes due to feeling unsafe (Congdon, 2015), and others describe a lack of boundaries, which can impact the relationships with participants (Thurairajah, 2019) and also mean that they take worries about their research into their domestic space. Low pay compared to some other jobs is also a source of concern.

Researchers contemplating a move into full-time roles as researchers can be put off by the lack of variation in the exclusive focus on research. Whilst they enjoy doing research, many also enjoy teaching and professional practice. As well as a reluctance to give these up, they worry that if all their responsibilities are focused on research, they will miss opportunities to think differently whilst carrying out other tasks and miss the structure that such jobs can give.

One of the tasks associated with being a researcher is to apply for research funding. Whilst some people enjoy the challenge of competing for money to carry out a project that is important to them (or will enhance their promotion application), others find that the time spent on applications is not only time away from research practice but also carries an inherent anxiety that it may all be in vain. Some of the administration and requirements to attend meetings that may be more or less relevant to their research are another reason researchers do not enjoy the role.

Although many researchers feel well supported by colleagues, others describe frustration when no one knows how to solve the research problem and that this can lead to feeling unsupported and alone. Although this is a somewhat unavoidable issue if the research is innovative and new, researchers often need to talk through their research and its challenges to rekindle their interest in and enjoyment of what they are doing.

Overall, however, it is important to remember that enjoyment, and lack of enjoyment, of research often fluctuate and to remind yourself of this when you question your enjoyment of being a researcher. In the next section, we will look at some resources available to enable and enhance success and enjoyment of researchers.

Staying abreast of research

It is important to stay abreast of research developments so that you can develop research ideas that will add to them. Technology makes it relatively easy to search for journal articles using databases such as Google Scholar, PubMed and the Web of Science. However, with the many sources available at a touch of a button, there is also a risk of information overload. It can mean that you have access to so many papers and articles that you can't read them all or that you get distracted and end up following avenues that are less relevant to your own research interests. A blog written for London School of Economics by Professor Anne-Wil Harzing (https://blogs.lse.ac.uk/impactofsocialsciences/2018/05/18/how-to-keep-up-to-date-with-the-literature-but-avoid-information-overload/) gives some useful tips on minimizing information overload.

Professor Harzing highlights that research is rarely so rapid that daily updates via social media are necessary. Instead, periodically review key journals in your field, looking up those of interest to see what they have published since your last review. If you have access to a library to do this, then articles of interest can be downloaded for later reading (although then make sure you do read them!). Google Scholar can also be used this way, after setting up your own Google Scholar citation profile page, allowing you to scan through your 'My Updates' page monthly to see articles that Google thinks will be of interest to you. The choice is based on your own publication history and articles you have saved to your Library, and whilst all may not be of the greatest relevance to your research, many are likely to be of interest. Harzing also suggests using technology to set up alerts for citations in new publications of your own works and for new articles in your field of research. This can be done on Google Scholar as well as ResearchGate.net and Academia.edu. She advises against being alerted when authors you follow have new citations, rather to be alerted when they publish new work. The Current Contents Connect feature of the Web of Science will allow you to request Table of Contents of journals as they are published, so a careful search input of journals most likely to publish in your field can yield helpful and quickly read alerts.

In addition to reading about new research in your field, you can, and should, talk to other researchers. This can be done by joining or setting up online networks, attending conferences and accessing online fora that discuss your area of research. There are many online networks that enable researchers with common interests to be put in touch with each other (e.g. Ireland Network for Pluralism in Qualitative Research, inpqr.net; European Association for Qualitative Researchers in Psychology, equipsy.org). You can also set up your own network, but be aware of the additional non-research-related time and work that are needed to keep your website up to date and interactive.

Conferences are often an enjoyable and fruitful way of attracting attention and feedback on your research as well as meet other researchers in your field. Don't be shy of contacting an attendee in advance to request a short discussion at the conference with them about their work or to approach someone whose research you are interested in and who you spot during the event. Most people are happy to talk about their research and will be pleased to discuss it with you. There are many online fora, and a short Google search for your topic is likely to find one. You may find fora not aimed at researchers (such as parenting fora) but which might include updates on recent research or calls for participation. PhD students have a plethora of fora to choose from, one of the most popular being thesiswhisperer.com. Checking in with these can provide a useful distraction from the intensity of reading through academic papers whilst also being educated on what it is going on in your field. You may also be able to contribute yourself, thus starting to establish your own name as a researcher.

Academic researchers who have some years of experience behind them are often able to apply for a sabbatical (commonly after seven years of service) or career break. Sabbaticals are meant to give academics time away from other duties in order to focus on writing and research. This can mean that you have more time than usual to really immerse yourself in current research whilst conducting and writing up your own. There can be great pleasure in this as you learn and have time to think about what else is going on in your research field.

It is also useful to subscribe to research updates issued by your own institution. All universities have a research office which offers many forms of support, from issuing grant application calls to supporting you in developing your own applications. Newsletters about completed and ongoing research within the university are regularly produced too, often with links to media presentations of the research.

Finally, the upsurge of ChatGPT can be helpful in keeping abreast of research. Whilst this is not a reliable source of information about research, it can be useful for brainstorming ideas about your topic that you can follow up and use in your key terms for a literature search on research databases. It can also simplify and summarize ideas that come up in papers you are reading, perhaps making them more accessible to you. Of course, such summaries should not be used verbatim in academic writing but as a way of considering and searching for ideas. Used appropriately, it can prompt your thinking and direction of your own research. For more ideas on using ChatGPT to keep abreast of research, see https://thesiswhisperer.com/2023/05/02/usingchatgpt/.

Public resources for researchers

Not every researcher has access to university libraries and other institutional support, so in this section, we will take a look at other available resources.

A quick Google of 'best podcasts for researchers' will give you a wealth of options. There are academic podcasts that range from podcasts presented by professors, lecturers and PhD candidates (e.g. 'In Depth, Out Loud') to podcasts promoting critical thinking, reasoning and public understanding of science (e.g. 'The Skeptics' Guide to the Universe (science)').There are also podcasts providing tips for lecturing (e.g. 'Lecture Breakers'), and in addition to those specifically for PhD candidates (e.g. 'Hello PhD'), there are podcasts providing career advice for ECRs (e.g. 'Behind the Microscope'). Others provide support for the well-being of researchers (e.g. 'The Self-Compassionate Professor') and on academic writing skills (e.g. 'Academic Writing Amplified').

If you prefer watching to listening as a form of learning, then YouTube might be helpful. Using the search term 'being a researcher', you will have a wide choice of videos presented by researchers and academics. These range from career development to confidence development, along with many videos on different methods of data analysis.

TED talks are another source of research information, again ranging from improving curiosity and research skills to foci on specific areas of research.

These free and easily available resources not only help you to gain new insight into queries and challenges that you are facing in your research but also might give you new ideas about research avenues to pursue. Above all, they may encourage your excitement about being a researcher by seeing and hearing how much it is valued and how important it is to explore.

Chapter summary

In this chapter, we have discussed some of the challenges of being a researcher whilst also having other responsibilities in academic, practice-based and community-based settings, and some of the ways these can be overcome. We have discussed the impact of personal characteristics, experiences of being frustrated and what it means to enjoy and not enjoy being a researcher. Different forms of support for researchers in all settings have been identified. Overall, we have started to explore how to be a researcher, and stay a researcher, against the backgrounds of competing demands and accountability to others.

In the next three chapters, we will focus in depth on each of the three researcher settings: Being an Academic Researcher, Being a Practitioner-Researcher, and Being a Community-Based Researcher.

Further reading

Edwards, A., & Talbot, R. (2014). *The hard-pressed researcher: A research handbook for the caring professions*. Routledge.

This book offers an introduction to experimental research, survey work, case study, interpretative research and action research for practitioner-researchers in the fields of health, education and social care.

Hook, H. N., Davis, D. E., & Van Tongeren, D. R. (2023). *The complete researcher: A practical guide for graduate students and early career professionals*. American Psychological Association.

This book covers character traits and skills necessary to become an effective researcher, describes the main steps for designing, conducting and completing a research project and discusses important considerations for building a career and research program.

Reed, M. (2017). *The productive researcher*. Fast Track Impact.

This audiobook shows researchers how they can become more productive by drawing on interviews with some of the world's highest-performing researchers, the literature and the author's own experience. It identifies important insights that can transform how researchers work.

Part 2

Researchers in Different Settings

4 Being a Researcher in an Academic Setting

Introduction

This chapter takes a detailed look at being an academic researcher. It describes the journey towards establishment as a researcher and focuses on some of the key milestones. It then discusses writing your research in different ways for different audiences, including for assessment, journal, conferences and funding applications. It concludes by exploring challenges and benefits unique to working as a solo academic researcher and working as part of a research team.

From undergraduate to early career researcher

Undergraduate level

The pathways to becoming an ECR in academic settings are fairly clear-cut. You are likely to learn about research and how to do it from the start of your higher education career, and as your studies progress, you will carry out research and learn how to use and consider research (your own and other people's) in how you read, develop and apply knowledge. The requirement for students to produce a research study for assessment frequently forms part of the final year and is often a more heavily weighted component of the class of degree awarded because research is such an important part of how academic knowledge is gained. Topics of study for research may be left to the student's choice – giving you an opportunity to pursue interests in areas of most significance to you – but are sometimes allocated. At this stage, the research is carried out alone or as part of a group and can be theoretical or empirical, depending on individual university structures.

Successful completion of a research project at undergraduate level is important in gaining the degree and it also demonstrates to future employers and educators your areas of interest and ability to complete a task.

Master's level

At master's level, the focus on research intensifies. Offering opportunities to develop skills as a researcher usually requires students to work increasingly independently. The time allowed often spans the duration of the programme at this level. A master by research (MRes) is dedicated exclusively to conducting research, under the guidance of a supervisor. Many people undertaking these programmes are doing so in preparation for applying for a PhD, so the choice of topic (and, sometimes, supervisor) is often related to the area in which they wish to develop a more in-depth study.

Having completed a master's level research dissertation, you will be well equipped to know whether you wish to continue and develop your research practice. The next step in the academic journey is to pursue research at doctoral level. You can pursue an academic PhD, a professional doctorate (DProf) or a PhD by publication.

Doctoral level

Academic PhDs are typically expected to take three to four years if pursued full-time or six to seven years part-time. Expected word counts are usually between 80,000 to 100,000 words and the study must make an original contribution to knowledge. The originality can be in topic (although if proposing to study an area not researched at all, be prepared to justify why this is important and why you are the person who should begin to explore this gap), participant group, culture or methodology. Whatever the perspective you wish to take, a clear case has to be made that there is a rationale for the study and that the study is worthwhile and feasible. There are also opportunities to research a topic defined by a supervisor – often because they have a laboratory of researchers already investigating the area and wish to add new ones or because they have attracted funding from external funders to recruit doctoral candidates to assist with new research projects. Occasionally, funding becomes available from within universities for supervisors to recruit new doctoral candidates, with offers of a stipend and a fee waiver.

Professional doctorates in academic settings differ from academic doctorates in aim and practice. Designed primarily for professionals (or aspiring professionals) to increase knowledge gained from working practice and apply it to their work, the doctoral research is usually part of a wider programme that seeks to advance the profession. The focus is on developing research that can be applicable to practice. This means that many professional doctoral candidates have the opportunity to investigate in-depth a burning question that may have arisen for them in the course of their practice. The thesis is of lower word count than academic theses, commonly between 40,000 to 45,000 words, to reflect the many other expectations of supervised professional practice, supervision and required course assessments. Other than that, however, the format is broadly the same as that of academic doctorates in that you are assigned a supervisor and expected to produce a thesis of doctoral standard. As with academic doctorates, the assessment is usually by viva voce in which a panel of experts conducts a verbal assessment of the thesis with you to ascertain that you have not only carried out but understood the research conduct and the outcomes that it has led to. The viva can be one of the most nerve-wracking aspects of pursuing a doctoral degree, so it will be discussed in detail later in this chapter.

A PhD by publication (or thesis by publication) is a doctoral thesis made up of published works or works aiming for publication, accompanied by a narrative text that explains and evaluates how the works form a coherent project. The works are usually expected to be peer-reviewed papers or books/book chapters. There are two routes to a PhD by publication: the prospective route requires candidates to write a series of articles (usually between four and six depending on discipline and institution), whilst the retrospective route (sometimes called a PhD by published qorks) requires candidates with existing research publications to collate and critically evaluate them. PhDs by publication can follow different models. The sandwich model (Gustavii, 2012) includes the articles as separate chapters framed by substantial Introduction and Conclusion chapters. The two-part model (Nygaard & Solli, 2020) sections off the works (or summaries of the works) and places them after the narrative. Both routes and the models of PhD by publication require demonstration of doctoral level scholarly knowledge, originality, publishability, critical thinking and research competence. Like PhDs and DProfs, they are assessed in both their written form and a viva voce. PhDs by publication are valuable ways of developing skills and a qualification required to enter academia but can also present challenges, such as navigating the hurdles of publication (including time delays and managing rejection), minimizing overlap in the works included and bringing together a thesis that has been developed in publishable parts (Paltridge & Starfield, 2023).

Requirements for PhDs, DProfs and PhDs by publication differ across institutions and countries, so it is always worth checking what is expected of you in applying for and completing your chosen programme. Many countries will only consider PhD and DProf candidates who have a master's degree, whilst some will consider work experience following an undergraduate degree as evidence of ability and commitment. Some require a master by research degree to be pursued first. The time in which the PhD is expected to be completed also varies by institution and country, from minimum periods of three or four years to a maximum of ten or eleven years. Most institutions make allowance for mitigating circumstances by allowing interruptions in study, or extensions of submission dates, but there can come a point at which universities begin to question the feasibility of the study being completed and remaining an original contribution to knowledge. Some universities consider allowing you to exit the programme at MPhil level, which is usually a point at which the study is shorter or less in-depth than a full PhD study. Assessment can be by a small panel or by your PhD committee and a public event in which you are questioned by a larger and more diverse audience.

Funding and support

In general (although check), fees for doctoral studies are less than those for undergraduate- and master's-level studies, so some people are prepared to fund themselves, perhaps whilst also earning an income within the university as a graduate teaching assistant (GTA) or a research assistant (RA).

Being a GTA or RA provides useful experience in developing as a researcher; being part of the academic environment and working alongside lecturers who are also researchers enables access to expertise on an informal basis. Teaching students is always a learning opportunity as you draw on your research practice to answer their questions about topics, methods, knowledge generation and acquisition and your own research.

Being an RA requires helping a senior researcher to conduct research by carrying out tasks such as data collection and analysis, and writing up research. This can be very helpful in starting to lay a track record of experience and publications which will further enhance your CV on completion of your doctorate.

Increasingly, universities are developing doctoral schools. These are designed to support you through your doctoral studies in ways other than academic supervision and consist of a range of staff, from academic to administrative, to advise and guide you through the expectations and requirements of the programme. They offer pastoral support, training, social events and conferences at which you can learn to present your burgeoning research.

Postdoctoral level

Once a doctorate is achieved, you may be interested in taking up a research fellowship at a university. This is a paid position in which you work under the guidance of senior researchers, either with them or independently, as a step in the development of a professional career as a researcher. There are many opportunities to hone skills and develop your research expertise, as well as to learn about different research perspectives. As a research fellow, you are also likely to learn how to apply for research funding and how to be part of a research team, perhaps even leading one. Often there are opportunities to liaise with others within and outside the university setting and to work on generating research publications. You may be able to earn a full salary, but the role is sometimes dependent on a stipend and/or generating funding from external sources. The expectations and range of abilities that research fellows have are wide, so you often must demonstrate not only research skills and understanding but also the ability to communicate and support others. You may be asked to teach doctoral or other student researchers and are likely to learn to develop high quality academic writing skills and an ability to generate new cutting-edge ideas for research.

Some people are happy to maintain their focus on research as a research fellow, often progressing to senior research fellow. Others like to combine their research with teaching practice and go on to become lecturers who can use their research to inform their practice. Researcher pathways can diversify and develop at this stage.

The next section focusses briefly on the dreaded viva voce and considers what it is as well as how to prepare for it and be as confident as you can when the time comes.

The viva

The viva voce is the requirement to defend your thesis 'with the living voice'. Sometimes called the public defence, the viva is a compulsory requirement of most doctoral programmes. It has to be passed in order for you to be awarded the title of Doctor. As the name suggests, viva voce is an oral assessment of your research and your role in it as the researcher. It is usually conducted by a panel of examiners who ask you questions about your thesis. The examiners will each have read the thesis separately and provided separate reports on it which they share and discuss with each other before meeting with you. The key purposes of the viva are to ensure the work is original and was conducted by you. This can mean they ask for more detail about sections of the thesis, about perceived omissions from it, and/or about issues within it which they disagree with. One of the more enjoyable aspects of vivas is that examiners can also discuss with you the aspects of the thesis that they have found particularly innovative or interesting. Often there is no set length of time for an academic viva so that examiners have the opportunity to inquire fully into areas they wish to know more about. Candidates are allowed to request a pause at any point, and often, a break option after each hour is built in.

Doctoral candidates can be daunted by the prospect of the viva, but believe it or not, many subsequently report having enjoyed the process. It is an opportunity to discuss your work with experts who will have read the thesis in detail and who can add some further thoughts to it, which in turn can help with the subsequent publication of papers from it.

Candidates are told the outcome of the viva immediately after it has taken place, following a private discussion between the examiners. Outcomes can be:

- Pass
- Minor amendments (such as adding short paragraphs, providing more detail in sections or correcting typos)
- Major amendments (such as re-development of particular chapters) which require more substantial work (and sometimes a second viva) to be done before the thesis is deemed to be of doctoral standard.

It is rare for a thesis to be failed because supervisors will advise candidates if they think it is not ready for submission.

A post-viva report is prepared by the examiners, detailing the requested amendments (which will have been discussed in the viva). If amendments are required, a time period for them to be completed is given (usually between three and eighteen months) and candidates are asked to re-submit the thesis once these have been attended to. One or more of the examiners will read these and give approval for the work to be passed or request for more work to be done on them.

Preparing for the viva

Given the importance of the viva, it is essential to prepare well for it. This preparation includes the selection of appropriate potential examiners by your supervisor(s). Examiners will be research experts and must have had no previous involvement with your work. The examiners will be drawn from outside the examining university (external examiner) and from within it (internal examiner). A chair will also be appointed to ensure that the process is fair and in accordance with university regulations and policies. Your supervisor will lead the selection and invitation process, so make sure that you have a full discussion with them before agreeing that they go ahead and contact the people you both think would be the best for the examination process. It is not enough to find someone in the field of your research. As you will have learnt by the time you get to this stage, there are many academic perspectives on the same topic, so finding people who have similar perspectives to the ones you have brought will enhance the opportunity for valuable insight and understanding of your research and how it was carried out and theorized. This often means that the examiners' work has been cited in your thesis.

Of course, by the time of the viva, you will be very familiar with your work, but it is always useful to think in advance about what questions might be asked of you. There are many websites with tips on this (e.g. thesavvyscientist.com, thePhDproofreaders.com), and it can be useful to talk to fellow students who have completed their vivas. Having a mock viva in advance with your supervisor can help you to anticipate questions and practise answering them in a succinct and knowledgeable way.

Anxiety can be high before and during a viva, so it is valuable for you to know that you can ask for a break or a moment to think about your responses at any time during it. It is also better to be honest in your answers if you are unsure about them. Examiners will help explore the area and perhaps advise that you research them further and add the information as an amendment.

Examiners know that candidates can feel under pressure and most are happy to help you with some introductory questions about your work that are designed to put you at ease (e.g. can you summarize your work for us in three minutes? What do you think are the strengths of your thesis?). The ensuing, more detailed focus on sections or chapters can then be tackled from a more relaxed stance (e.g. why/how did you choose to include these theories in your literature review? What other methods did you consider before deciding on this one?).

General structure of PhD viva questioning

- One or two opening questions
- Questions focusing on the research aims, objectives, questions and hypotheses, and how they were reached
- Questions focusing on each chapter of the thesis and on particular aspects of individual chapters
- Questions about challenges in the research and how they were met
- Questions about the contribution of the research and the implications of the research outcomes
- Questions about whether and how you would have done the research differently and future research avenues that have emerged from your research

Overall, perhaps the best advice for a successful viva is to re-familarize your-self with your research (vivas can take place a long time after the submission of your thesis depending on the examiners' availability), anticipate what you may be asked about and use any techniques you have to enable you to stay calm and focused during it. The successful outcome will make the process very worthwhile as you celebrate your achievement!

Writing skills for academic researchers

As we have seen, research is much more than collecting and analyzing data. A key skill is to be able to write it up for others to read and benefit from it. Furthermore, the writing-up process in itself can help to crystallize ideas and understandings of your research and, therefore, makes learning the skill of academic writing the key to a successful researcher. Writing up can be for assessment, as with dissertations and theses, and for dissemination through journal publications, reports for conferences and other audiences and presentation to communities. Each requires different styles.

In general, academic writing aims to convey ideas and arguments to scholarly audiences. Even if these are complex, the writing should not be. Instead, it must show how the ideas and arguments develop, using references and citations of others to illustrate and justify their foundations. The language is precise, and – in some disciplines – technical, with the aim of explaining, describing and critiquing your own work and that of others. Some key characteristics of academic writing are summarized below.

Table 1 Key aspects of academic writing

Characteristic	Writing style and purpose
Is clear and concise	Succinct and to the point. Avoids jargon and excessive wordiness.
Signposts	Guides the reader. Shows connections between ideas and transitions from one point to the next.
Provides evidence	Uses appropriate references and other credible sources to support arguments. Shows how data is transformed into results.
Is original	Summarizes, paraphrases or re-presents other writing to show researcher interpretation of its meaning and relevance.
Has a formal tone	Avoids contractions, colloquialisms, personal opinions and informal expressions.

Academic writing for assessment

Writing up research for academic assessment means adhering to the university's expectations of content and structure. Typically, the structure of a dissertation or thesis requires a literature review, methodology, results/findings/analysis and a discussion. Evidence of ethical approval and process (consent and debriefing forms, for example), recruitment posters or social media messages and a sample of data and its analysis may also be expected. Sometimes, a portfolio consisting of a context statement, thesis, case study and publishable paper is required for assessment, particularly in professional doctorates.

The style of writing will vary by discipline and research approach. Theses that present scientific research will be written in an objective, third-person style whilst social sciences and humanities are more likely to be written using the first person to show the researcher presence in the research.

Learning to write for a thesis or dissertation can be challenging. In addition to the expected format of the work, the word count can feel limiting, requiring succinct and clear presentation to ensure sufficient depth and detail. It can take practice, and some researchers (and their supervisors) like to review the draft chapter by chapter whilst others prefer to read a complete draft. It is important to discuss this with your supervisor so that you agree on how you will receive feedback on your drafts.

It can be hard to have your carefully crafted work critiqued by your supervisor, but it can also be wise to be open to their feedback and comments. Even if what you have written is clear to you, if it isn't to your readers, then it needs development. Supervisors are experts in knowing how research should be written up and theses presented and will read them with an eye to their assessment by examiners. Although you can submit your thesis whenever you think it is ready, it is not advisable to go against the advice of your supervisor. They will work with you to reach a point when it is ready, even if this means completing more drafts. Working closely with them will minimize the possibility of examiners asking for major amendments at the viva stage – although always be prepared for examiners to highlight omissions or request further detail to be added in particular sections. It is helpful to see this as a way of strengthening your work before it becomes available to a wider audience, and it prepares you well for journal article reviews.

Academic writing for journals

A common mistake made by newly graduated doctoral researchers is to cut and paste the thesis to fit the shorter word count of journal articles. This often results in a rather clunky manuscript that can over-focus on certain aspects of the research and not provide enough detail of others. Journal articles are another craft in themselves, and editors and reviewers are experts not only in the topic but in the style of articles for their journals (see e.g. Finlay, 2020). Many journals have editorials detailing the style and detail they expect to see in submitted manuscripts, and it is always worth reading these as well as the aims and scope of the journal and information about manuscript formatting, word count, etc. before writing the article.

Key to having a manuscript accepted by a journal is ensuring that it meets the criteria and focus of the journal. It is no good submitting a methodological paper to a topic-focused journal and vice versa. It is also important to be prepared to omit some points from your journal article in order to highlight a key focus of the research and its outcome. A useful rule of thumb is to ensure that the paper centres on one aspect of the research. This might be method, a key finding or the innovation of the research process.

If not rejected at the editor's desk because the content does not fit the journal's aims and scope, the manuscript is sent out for review to two or three experts. Although many journals aim for this process to be as quick as possible, they are dependent on the reviewers' availability and willingness to review, so be prepared to wait a while before hearing back from the journal. The review process is 'blind', so ensure that you have appropriately anonymized your submission.

Once the paper is returned to you by the editor with their summary of the reviews and the reviewers' reports, you are given a time period in which to make the amendments before re-submitting. It is important to remember that reviewers' suggestions are given with the goal of strengthening the paper and preparing it for the readers of the journal, but if there are some comments you do not agree with, you can explain why in a letter you send back to the editor along with the amended paper, showing how you have addressed (or not) the reviewers' comments. The letter should be respectful and polite and provide clear responses to each reviewer comment, explaining where you have made the suggested changes or why you have chosen not to.

The reviewers are sent the amended paper and give their decision about its readiness for publication. This process is usually faster than the initial review because the paper is already known to them and they will be looking primarily to see how their comments have been attended to.

You will then be notified of the outcome of this process – further amendments, acceptance or, occasionally, a rejection.

The process of writing and submitting manuscripts for publication can be a nerve-racking one for new researchers. Obviously, you will be keen to have your work accepted and a rejection can feel hurtful. Ensuring the manuscript is appropriate for the journal and well written with as few typos or other distracting features in the writing as possible enhances the chances of its acceptance.

As all writers know, it is helpful to read as much as possible to aid your writing. Reading journal articles helps your thinking as well as your writing and gives insight into the different styles and foci of different journals. It is helpful to become a reviewer for journals you have an interest in. By asking questions of other submissions and crafting constructive feedback for authors, you will develop your own skills as a journal article writer too.

The following reflection illustrates some of the mixed emotions that can be experienced when doing academic writing.

Researcher reflection (from an experienced researcher and academic writer)

It is like doing a literature review when engaged in research. Yes, it has its tedious side. But I like the stimulation of the new learning. I also like playing 'detective' and following up on clues and elusive references. It's satisfying to follow up on little clues like the odd reference in an article, to discover a new piece of 'evidence'.

Writing research grant applications

Applications for funding are competitive, and each grant will have its own requirements for the structure, presentation and evaluation of the research proposal. They comprise a mixture of writing with academic detail and in lay terms, and showing budgets in detail and in overview. Grant applications are, therefore, quite different in writing style to research reports or journal articles and will require you to draw on a myriad of skills. Furthermore, grant applications may be read and assessed by a diverse, often multidisciplinary, audience comprised of experts in the field of the proposed research, experts with the knowledge of related fields, financial experts and experts in the use and application of the research focus. There may also be readers who are trustees of the funding body. Each member of the reviewing panel must be convinced that your application shows that the research will meet the funder's objectives and priorities, and bearing in mind that there may be hundreds of applications for funding from the same funder, a clear and compelling case for yours to be the chosen one is essential.

The writing must adhere to the word count (or sometimes characters, in which punctuation marks also count) of each section, and the writer must strive to complete each section by answering the questions in it as clearly and succinctly as possible. Overall, the application needs to be as persuasive as possible, so that its assessors are convinced not only of the feasibility of the study but also of the good use of their money, and the researcher's ability to carry out the research well and within the timeframe.

Writing research grant applications is, therefore, a skill that must be learned and honed in order to enhance the likelihood of success. There are courses available to help researchers learn how to write grant applications (e.g. masterclasses.nature.com), and funding bodies provide guidance on how to complete the application, but many researchers say the best way to learn is by working with colleagues who have been successful in their applications for research grants. This helps develop knowledge and skills from their writing and perhaps by co-writing or drafting individual sections for feedback. You will come to recognize the information required in different sections and learn how to present it in succinct and powerful ways.

Having considered the different writing skills needed to be a successful academic researcher, the next section now turns to disseminating research through conference presentation.

Presenting at conferences

Presenting your research at conferences is a valuable way of attracting attention to it, holding discussions about it and inviting comment and constructive criticism on it. Conferences also provide networking opportunities and enable you to learn from other researchers. Furthermore, conferences usually have social events at which you can reconnect with other researchers and meet new researchers in informal settings. Another bonus is that there are often conference discounts on books, which not only keep you up to date on the latest publications but also enable you to buy books that you may not otherwise be able to afford.

However, presenting at a conference can be a daunting experience for even the most experienced of researchers. It requires confidence (or at least the appearance of confidence) in public speaking, thorough knowledge of your research and a preparedness to answer a range of questions about your work. With the onset of the COVID-19 pandemic, many conferences had to be held online, and it looks like many will continue with both online and in-person presentations for the foreseeable future, requiring additional skills in slides' preparation and delivery. In this section, we will consider some ways to prepare for conference presentations and ways to ensure that yours is interesting and engaging to the audience.

Perhaps, the most important consideration is to carefully select the conference you wish to present at. There are a wide range to choose from, particularly during the busy conference seasons when teaching has ended for the year (generally June to September in Europe) and they are held in a number of national and international locations. Finding out about upcoming conferences can be hard but signing up to university research office lists, regulatory bodies for your discipline and national networks is often a good starting point. Let your supervisor or team members know that you are interested in presenting at a conference, and use your own network contacts too. If there is a particular conference you have previously heard about, then look for it online and see when the next one is scheduled. Look out for calls for papers (CfP); society bulletins and journals can be useful sources for this.

It can be tempting to decide on a conference because it is held in sunny climes or in a country you have always wanted to visit, only to arrive and find that there is little interest in your paper or few people with shared knowledge of your research topic or approach. Whilst it is always useful to gain experience in presenting at conferences, it can be dispiriting to have spent time and money in presenting only to come away feeling that little insight to your research has been gained.

In addition to considering the focus of the conference, it can be useful to think who and how many other people will be there. Starting your conference delivery career at a large, international audience can be overwhelming and contrasts greatly with smaller conferences that aim to attract doctoral and ECRs in addition to those who are more experienced. The interest and questions are likely to be more helpful and more freely given, and other researchers you meet there can become supportive colleagues when you each return to your place of study or work.

Sometimes, supervisors are happy to attend conferences with you. This can be very supportive, guaranteeing you at least one member of the audience, and with some pre-planning, a question you are pleased to answer. You will be introduced to colleagues of your supervisor, and their students, thus starting your own network of contacts.

Once the conference you wish to attend has been decided on (and you have considered the feasibility of the conference fee, travel and accommodation, either as a self-funding attendee or as agreed with the university), it is time to write the abstract. Writing an abstract for a conference means describing your research and the focus of the paper you wish to present. Abstracts have a limited number of words, often between 250 and 500, so your skills of writing clearly and concisely will need to come to the fore. Often CfPs come out six to nine months prior to the conference date, so it can be tempting to (over)predict the stage and the results of your current research. There are many tips and guides online to writing a high-quality abstract (e.g. https://www.ncbi.nlm.nih.gov/pmc/articles/PMC3732725/) but in short, make sure it includes all the information requested (usually a title, hypothesis/research question, methods, results/findings, implications and conclusions), adheres to the word count, and, above all, conveys the relevance of your research. Remember that not all those reading the abstract will be familiar with the topic or methods of your research, so make sure the details are conveyed clearly. Finally, although writing 250–500 words sounds like a quick and easy task, it can take a lot longer than people anticipate. It can be hard to know what to omit and equally hard to ensure that the key point is adequately made. Expect to re-draft it before submission, and seek feedback from a fellow researcher or your supervisor. Always check and double-check for typos and formatting.

As the conference draws near, preparations for delivering the paper can be started. Again, don't underestimate the time this process can take. Conference presentations are typically between 15 and 20 minutes, allowing 5–10 minutes for questions afterwards. It is bad manners and frustrating for the chair of your session if you run over time, takes away your opportunity to discuss the research with the audience and runs the risk of you being cut off before the paper is finished. If you are including slides with your talk, these need to be prepared along with the talk.

The best slides are usually uncluttered, with a few words and/or a simple figure to summarize the focus of what you are saying. When presenting in person, the slides should not be too animated or detailed – you want the attention to be on what you are saying and not on the slide. Conversely, when delivering online, the attention will be on the slides, so they can contain a little more information and a bit more animated entertainment. The best conference presenters do not read from the slides but use them to supplement and illustrate what they are saying. This not only holds the audience attention more but also enables your passion and excitement of the research to be better conveyed. If the audience is engaged with you, they will be more likely to ask questions and comment when your paper is finished. A rough rule of thumb for slides for an in-person conference presentation is one every two minutes. It may be shorter or longer for online slides.

When the time comes, make sure you have your talk and your slides safely stored in your head, on memory sticks and on your computer. Find out where the room you will be presenting in is (or what the link is) and make sure you arrive early for the session. It is bad manners to turn up halfway through a session just in time for your paper and means you cannot test the equipment, introduce yourself to the chair and get a feel for the room. Arriving early online allows you to meet the chair and other presenters as well as test the link. Anything you can do to reduce your nerves is valuable now. You can anticipate some of the questions that may be asked and how you will answer them. You can also prepare yourself for responding to trickier questions by thanking the questioner and offering to meet them after the session or over a coffee to discuss it further. Remember that most people attending a conference will be presenting too and know the anxiety (and value) of doing this and, so, will be happy to speak later to help develop your research. Rarely do people want to catch you out, so consider all questions as helpful.

A final note: it is a good learning opportunity (and sometimes required by universities) for you to make your first conference presentation in the form of a poster. Whilst the amount of work required is no less, the preparation is different in that you do not usually have to talk about it to the whole audience, instead engaging interested individuals as they look at it. It is just as important to have the relevant information on the poster, this time with the aim of attracting people to it. Make the title snappy and informative and, as with slides, do not over-clutter it. Use a mixture of words and illustrations. Containing different sections in boxes or using arrows to show a clear journey through the research process helps those looking at it to follow your decisions, actions and results. As with verbal presentations, remember to acknowledge supervisors and any other researchers involved in the research.

Reflective question

What experience do you have of public speaking? What are the similarities and differences to presenting at a conference? How can you use any experience you have to prepare for giving a conference presentation?

In the preceding sections, we have focused on dissemination of research through writing and conference presentation. In the concluding sections of this chapter, we will consider some of the social and communication skills that are necessary to academic researchers, starting with being a solo researcher.

Social and communication skills

The solo academic researcher

Carrying out your research alone is usually a key requirement of master's- and doctoral-level research to ensure that you have developed the necessary research skills to reach qualification.

Working on your research as a solo researcher has many benefits: you can set your own timetable, take time to think through challenging aspects of it, make decisions and determine its direction as it unfolds. However, it can also, at times, feel a little isolated and can threaten your motivation. There are many ways you can counteract this.

Many universities have PhD student offices and labs where you can go to work alongside other PhD students. This not only gives you opportunities to discuss your research informally but also to pick up tips and advice from fellow students. Working there regularly gives you 'somewhere to go' other than your home and the chance to strike up friendships and working relationships. This can be particularly important when you are finding the research stressful or frustrating, but is also a great opportunity to celebrate successes. Some PhD offices have a 'Wall of Fame' on which students post their publications, conference presentations or anything else they are proud of in their research. This creates a supportive and encouraging atmosphere in the office.

Recognizing the importance of supporting PhD students means that some universities have an external group facilitator they call on to offer regular, informal, confidential group sessions where non-academic matters can be discussed and shared. Students can set up their own peer support group as an alternative or a complement to these.

Working alone also means that you can take a break from the research if it becomes overwhelming, without worrying that it will affect other researchers. Although the pace of the research will often be set by the participants and the deadlines, it is wise to recognize times when you and the research might benefit from a break. This might be when you are feeling distressed by some of the personal experiences you are researching or because of analyzing sensitive data, for example. Taking time away, either completely or by turning to another aspect of the research that is less emotionally demanding (a literature review, for example) can allow you time to refresh yourself and return to the research with renewed vigour. It is always useful to keep a research journal, and this can be helpful when negative emotions seem to be provoked by the research. In addition to keeping an audit trail of processes and decisions in the research, a journal can include personal feelings and challenges which do not ever have to be publicized.

When researching alone, it is always invaluable to maintain good communication with your supervisor. These experienced researchers may well be able to recognize and help with 'stuckness' or emotional overwhelm that is affecting your research and can draw on their knowledge of ways to manage them. As a researcher, you should be able to feel that you can raise personal issues that are a distraction or hindrance to the research, without fear of judgement or criticism.

Being a researcher in a team

Being a researcher in a team may at first thought seem to be an easier option than being a researcher on your own. You will have ready access to fellow researchers, enjoy the prospect of working together towards a common goal and be able to share and discuss aspects of the research experience with others who may also be going through it. In practice, though, there are some times when team-working can bring challenges that are not encountered as a lone researcher.

A research team may be multidisciplinary, interdisciplinary or transdisciplinary (Slatin, Galizzi, Melillo & Mawn, 2004). Multidisciplinary teams consist of members who bring their specific disciplinary perspective to work on a commonly agreed research problem. Interdisciplinary teams are made up of members who agree on a dominant approach or mixture of approaches to an agreed research problem. Transdisciplinary teams bring together specific concepts, theories and approaches to develop and use as a shared conceptual framework. Each member of the research team will have skills and expertise to contribute. So, the first task will be to agree on the research focus and questions to be asked. Straightforward as this sounds, think about the differences in perspectives and approaches that can enrich but also challenge a team. Researcher biographies, interests and priorities will play a key role in deciding on the research problem and how it is to be researched, and researchers from different disciplines may understand the problem differently and with different theories. Those with expertise in different research approaches may argue for particular methods to be prioritized (commonly, debating whether qualitative methods should be secondary to quantitative methods). If the team is composed of researchers from academia, professional practice and other spheres, there may be different priorities and availability of time prioritized for research for each team member. Teams made up of researchers with different levels of experience in research may require additional mentoring of less experienced team members.

To address some of these challenges, it can be useful for each team member to be clear as to what expertise and knowledge they are bringing to the team and to be prepared to learn about the expertise and knowledge of fellow team members. More experienced researchers can be reassuring about their willingness to support and teach less experienced team members, and less experienced team members can benefit from feeling able to ask for help when they need it. All members of the team should remember that the language they are accustomed to using in their discipline-specific research, or research approach, may not be familiar to other members of the team. Being clear in communication, willing to explain and feeling able to ask and answer questions greatly enhances the progress of the research and the coherence of the team of researchers.

Whilst many of these discussions can take place as the research is set up, ongoing regular reviews and discussions can maintain open dialogue and help the team to work together to address challenges as they arise in the research process.

Working as a researcher in a team also means becoming aware of one's own and others' personal characteristics. As a researcher, you may feel shy or anxious about appearing less competent than other team members. Conversely, you may feel frustrated at the perceived lower level of ability or research progress of other researchers. Adopting a reflexive stance when working as a team can help with this. Not only do you need to consider your engagement with the research as an individual but also your relationships with other researchers in the team and how these shape and inform the research. Personal reflection about your relationships with other team members can help a reflexive team-wide dialogue in which working and personal dynamics can be better understood. This can generate a valuable plurality of contributions to the research and discussions of it amongst the researchers involved. Better research usually evolves from developing strategies in response to openness to the knowledge that other researchers bring.

This approach can be helpful when the research methodology or understanding of the topic differs to that which you usually employ when working as a lone researcher. It allows for questions to emerge that can be answered by bringing the different perspectives of the team to it, and enables each team member to highlight their contribution. Being prepared to acknowledge your own biases and assumptions as a researcher goes a long way towards being prepared to hear other perspectives and minimizes the chances of gaps and apparent contradictions being overlooked or dismissed. The resulting research is likely to be richer and more nuanced as a result.

Assigning clear roles to each team member is also helpful to researchers working together. There needs to be an identified team leader (or an agreement to work collectively), someone to take responsibility as an administrator (e.g. organizing meetings and maintaining records of them, timetabling key tasks such as submissions to ethics committees and keeping track of expenses claims for participants) and someone who manages the budget, who could be the administrator or a separate treasurer (sometimes, it may be the team leader). Other roles can include expert participants, participant recruiters, data gatherers, analysts and a documenter (to keep a record of the progress of the research and ensure reports are written on time). With a clear team structure in place, each member knows who to approach with specific queries or concerns and will know their role in different aspects of running the research.

It is sensible to agree in advance on the plan for writing up the research. If the team is multidisciplinary, made up of practitioners and academic researchers, for example, the priority and time available to write up the research for publication may vary. Academic researchers are usually keen to have their work published whilst practitioner-researchers may think it more important that the research is disseminated in community events or short reports. The different aims for dissemination should be clarified as soon as possible so that expectations that are not feasible or desirable are not put on different members of the team. Similarly, when it comes to writing up, lead writers should be identified and roles for the other team members agreed (such as, perhaps, contributing a certain section of the manuscript or report, presenting it at a conference or proof reading). Finally, the order of authorship on submitted papers should be agreed to avoid any frustration or resentment when it comes to submitting the research to journals or funding bodies.

It can also be the case that you are researching in a different location to other team members. This can make working to maintain support structures and remaining in contact with the team even more valuable, not least because it will help to keep your research motivation up (Koehne, Shih & Olson, 2012). Whilst more formal team-based reviews may be timetabled into the research process, it can also be valuable to have informal connections with some or all of the team – perhaps a WhatsApp group or Zoom meetings to discuss personal aspects of doing the research.

Chapter summary

This chapter has described pathways to becoming an academic researcher, highlighting the various steps that must be taken and the assessments along the way. It has included a particular focus on the viva voce, including what to expect and how to prepare for it. It has discussed the various writing skills needed by academic researchers for assessment, publication in journals and for crafting research funding grant applications. Research dissemination through conference presentations has also been explored, including differences in presenting online and in-person. Finally, the chapter explored what it can mean to be an academic researcher working alone and as a member of a team. In Chapter 5 we turn to being a researcher in professional practice settings.

Further reading

Evans, D., Gruba, P., & Zobel, J. 2011). *How to write a better thesis* (3rd ed.). Melbourne University Publishing.
  Written by experienced research supervisors and trainers, this book takes an integrated, down-to-earth approach, drawing on case studies and examples, to provide step-by-step guidance towards productive success.

Freiermuth, M. R. (2023). *Academic conference presentations: A step-by-step guide.* Palgrave Macmillan.

This book takes readers from the initial idea for a conference presentation through the abstract submission, and to the presentation itself. Drawing on the author's own experiences, it highlights good and bad practices and provides tips on issues such as writing an abstract, negotiating group presentations and presenting at virtual or hybrid events.

Rainford, J., & Guccione, K. (2023). *Thriving in part-time doctoral study: Integrating work, life and research.* Routledge.

This practical guide helps doctoral researchers to navigate their learning experience alongside the work and life challenges they may be facing during their doctoral journey. It provides realistic advice, learning points and reflective activities based on real experiences.

5 Being a Researcher in a Practice Setting

Introduction

In this chapter, we explore what it is like to be a researcher who is also a practitioner (referred to as a practice-based researcher in this book). By practitioner, we mean someone who is delivering services or creating products for clients. They may be carrying out research in their own practice or elsewhere and on the services or products of their practice or on those of other practices. We start by considering what it means to be both providing services and products, and carrying out research. We explore how to balance boundaries and recognize limits and institutional expectations and then move on to discuss ways of finding time and money to do your research, before ending with a detailed look at disseminating research for different purposes.

Being a practice-based researcher

Being a practice-based researcher means that you conduct research in your workplace or about the services or products that you or your organization provides. Your workplace might be a healthcare setting, corporate organization, education service or industry. You may be a full-time researcher for your organization or carrying out research as part of your role as a practitioner. This can mean combining research with service delivery, development or monitoring.

Research conducted by practice-based researchers often has a more practical than academic focus and aims to produce outcomes that can be applied to bring about change. Desired changes may be in improving services, contributing to policy development, developing new products or monitoring competition. Therefore, the research is often solution-focused and used for organizational decision-making and strategy development. The impact of the research is usually evaluated by the measurable effect it has on services, policies and products. It may also be that it brings about increased understanding and awareness of a product, enhances health and well-being of clients or contributes to the economics of an organization by increasing savings or growing income.

Whilst the standards expected of the research conducted by practice-based researchers remain as high as those expected of academic and other researchers, there are some issues that require particular consideration by practice-based researchers.

In some institutions, research is secondary in importance to service delivery. It may be that demands made for the service mean that all available practitioners are expected to prioritize their time for service provision. Or it may be that there is sufficient confidence in the service or product that it is felt that no further research is necessary. As a practice-based researcher in such settings, it can be useful to highlight to colleagues how increased understanding and awareness of the service and developments in it can be beneficial to the institution. You can bring colleagues' and management's attention to existing research that has direct relevance to the improvement of services or products you are delivering. Staff meetings may be useful to initiate discussions about research and if you know of a researcher at another institution who has produced relevant research, you may be able to invite them to come and talk with colleagues about it. Perhaps, most effective in highlighting how useful research can be is to use it yourself and share the improved service outcomes.

The primary research goal of practice-based researchers is to have their research translated into practice. The research is usually contextualized by the institution where the researcher works, so this can mean researching with clients you deliver services to or with fellow practitioners providing the services. It can be challenging to recruit participants who are also colleagues, perhaps because they do not want their work scrutinized or because they are wary of change being brought about as a result of your research. Being a researcher who is seeking to change or evaluate services or products provided may bring you into to conflict with what has been called the 'shadow side' of organizations (Fox, Martin & Green, 2007).

In the shadow side are the issues that are covert, undiscussed, undiscussable or unmentionable in an organization (Egan, 1994). Think, for example, of introducing formal procedures for booking annual leave online compared to dropping into a line manager's office to ask for time off. Members of organizations may be happy in the knowledge and use of systems that they agree with and may not want research to expose and, possibly, change. On the plus side, however, you as an 'insider' practitioner-researcher are well placed to know about these systems and how they work. It is important to remember that the shadow side of organizations can also be beneficial to the way they work, so exposure of them may be welcomed by members of the organization so that they can be more formally implemented. An example may be that some colleagues work from home on some days, and this is a policy that can be formalized to allow all to utilize it if they wish.

Maximizing impact

Maximizing the impact of your research as a practice-based researcher is key to gaining support for it. Many research funders now ask for the 'pathway to impact' that your research and its outcomes will follow (e.g. the UK's National Institute for Health Research). Pathways to impact typically include inputs, activities, outputs, outcomes and impact. See Figure 1 for a summary.

Figure 1 Pathway to impact diagram (adapted from the WK Kellog Foundation Logic Model, 2004)

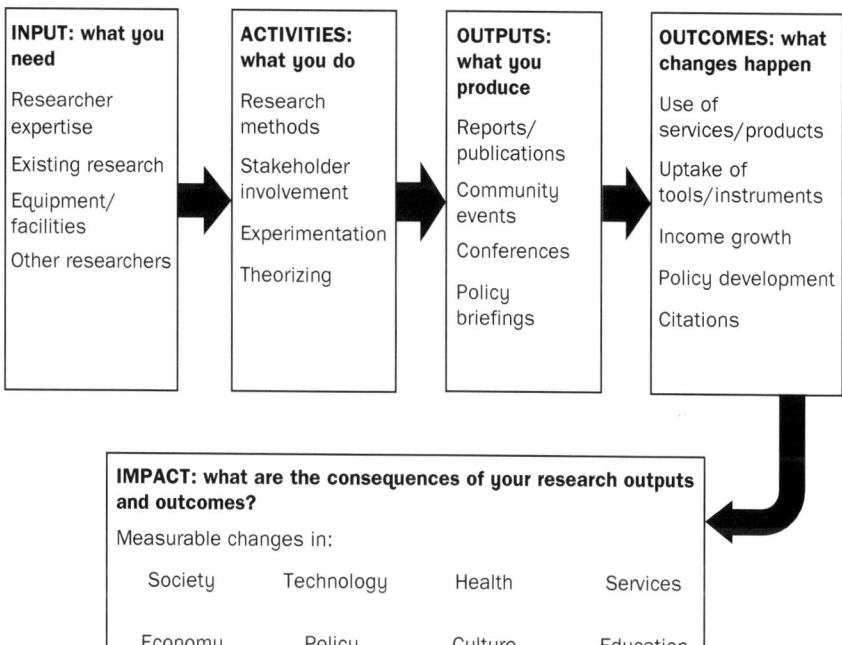

In a nutshell, practice-based researchers strive for the research outcomes and outputs to be integrated into practice. This means treading a fine line between ensuring the research is taken seriously and its findings considered with a view to implementation, and risking the research being ignored or, worse, suppressed. Clearly, to maximize impact, you need to consider it in the design stage of the research study. Is it an evaluation that can gather quantitative data to show outcomes of the services? Or is it a change process study that seeks to change services themselves? What evidence will be provided by the research for others to consider using it?

As the researcher, it will be useful to start the marketing of the research to colleagues and management as early as possible; letting them know the potential value of it and how its implementation can enhance their practice, work-lives or organizational development is something that can be done from the outset, and keeping colleagues up to date with the emerging evidence can bring them further on board. Similarly, if relevant, it can be important to make clear to others that you are carrying out the research as part of study or training elsewhere. Colleagues will know then that the assessment may be by another institution and that whilst they may be interested in it, its findings will be more generally focused on implications for the profession as a whole rather than for their own service delivery.

The following reflection highlights what can be gained by doing research as a practice-based researcher:

Researcher reflection (from an experienced practitioner with a recently completed doctorate)

I find that doing research helps me to connect more deeply with the areas of my work and the world that I feel so passionate about. It's a little haven within the bustle of clinical practice, where I can really explore the depths and dark corners of important and fascinating parts of human experience. It's a creative process more than anything, that enriches not only my aim to help others but also to grow within myself.

Collaboration

Some of the challenges to conducting research on your own as a practice-based researcher can be addressed by researching collaboratively. This may be with colleagues from within your organization or with those in other organizations (or both). It can be useful and of value to the research to bring in different research expertise, stakeholders with different agendas and interests in the research and clients for whom the research is meant to be beneficial. It can also help with your involvement in the research (because you may be more research-active at some stages than others), the likelihood of your research having a greater impact (because there is a wider dissemination network) and your skills-development (because you can learn from others about different ways of doing research).

Doing research collaboratively means you will be leading, or working as a member of, a research team. The team may be multidisciplinary, in which you work alongside other researchers to each bring a specific disciplinary perspective to an agreed research problem; interdisciplinary, in which you work on an agreed research problem using a dominant approach or an agreed mixture of approaches; or the team may be transdisciplinary, in which you work to develop a shared conceptual framework for an agreed problem using discipline-specific theories, concepts and approaches (Slatin, Galizzi, Melillo & Mawn, 2004).

You will be a member of this team because your research skills are recognized as important to the research. It is an opportunity to share and discuss your research expertise and identify areas of further development in them. It also gives you insight into and understanding of other perspectives on the problem, allowing for a more complex understanding and research outcomes that can be of use to a wider audience. You can expect a clear rationale for your inclusion in the team and, in turn, be accountable for providing a clear rationale for your contribution to the research. This can be particularly important when you are a practice-based researcher working with academic researchers because agendas and priorities in the research may differ. As a practice-based researcher, your ultimate research aim is to provide research-based evidence for use in services and products. Academic researchers may have more time for research than you do, and you may need to explain to them the time you have available for research. When working collaboratively with researchers who use different methods to you, you may need to ask questions about their methods and what perspectives they are bringing to the research so that you can fully understand the insights that may be gained from the research. Similarly, you should be prepared to explain the method you use, including how much time is needed for data analysis and interpretation, remembering that qualitative and quantitative methods typically have different timescales to allow for data transcription, analysis and interpretation. Not knowing this can lead to frustration at perceived delays in team-member contributions. Throughout the research process, you should strive to encourage open communication between team members by using accessible language, expressing interest in and curiosity about the research contributions from other members, and participating in regular reviews and update meetings. If you are the team leader, it is important to the success of the team to be open to and supportive of all team members, as well as ensuring that those looking to develop or extend their skills, or those with less research experience than others, can learn from being a member of the team.

Collaborating with others to do research also benefits from expectations of each team member being clear from the outset. There are different roles in research teams as well as different phases in the research. It may be that you are more comfortable analyzing data than collecting it or that you are more comfortable than other members in writing up or editing. By making these preferences and skill sets clear to other team members, and finding out where their preferences lie, the expectations of what work is to be done and when can be made clear for each member. Ideally, everyone will be assured that their preferences and skills lie within expectations, and a successful, efficient team will see the research through.

It is also useful to determine authorship of eventual write-ups of the research so that all team members feel that their contribution is recognized by a wider audience. Authorship can vary with the style and presentation of the write-up so that, for example, practice-based researchers can be the lead on reports for policy-makers and those responsible for strategy-development and academic researchers can take the lead on journal articles. Deciding authorship at an early stage can minimize disgruntlement or the feeling in team members that their contribution has been minimized or overlooked, both as the research progresses and when it has been completed and is presented to the world.

Finally, collaborations inevitably include power dynamics. Whether these are because you feel more or less knowledgeable about the research than other team members or that you feel your contribution is undervalued, it is important to recognize and address that differences in expectations and understandings can create power dynamics. Other team members who may not be familiar with your research approach or methods may not give it the same status as their own, for example, resulting in your perception that your work is not being recognized or valued. You may feel that you have been brought into the team to 'tick a box' or to act in a role that does not avail of your full skill set. Again, openness and transparency in communication can go a long way in addressing such issues before they become a source of resentment or disappointment, but it can also be useful to adopt, and encourage, a reflexive approach. Working collaboratively means that engagement with the research is not simply between you and the research but, instead, between you and other researchers' engagement with it. This can mean that you seek to understand and question the purpose and contribution of other methods or researcher roles and consider how they apply to the research study as a whole. It also means that your own consideration of the aspect of the research that you are conducting is clear to you and accessible to other team members.

There is no doubt that researching collaboratively can be of benefit to research as well as to you. It can bring diverse perspectives to the study, making it useful to a wider audience; it can help with more effective and relevant dissemination; and it can greatly benefit the development and extension of your research skills and confidence within time frames that enable you to be a practitioner as well as a researcher.

Ethical considerations of practice-based researchers

As a practice-based researcher, you will need to adhere to both the ethics of your practice and research ethics (see, for example, Ethical Guidelines for Educational Research of the British Educational Research Association, 2024). Practice-based researchers have special opportunities to make 'research ethics' 'everyday ethics' so that both their research and their practice can interplay and inform each other to bring issues such as informed consent, striving to do no harm and understanding power dynamics together in their professional practice and their research.

Mockler (2014) proposes five overarching ethical guidelines for practice-based research:

- Observing ethical protocols and processes
- Pursuing transparency throughout the processes
- Agreeing on a collaborative goal with participants
- Seeking transformation that leads to action in practice
- Justifying the research to its own community of practice.

Guided by this ethical framework, quality in practitioner research is enhanced in three key ways:

- Quality of evidence (relating to the processes used to elicit, gather and analyze evidence)
- Quality of purpose (the ways in which practitioner research is conceived and enacted within the practice environment)
- Quality of outcome (the balance of critical and positive stances taken towards the practitioner research).

Whilst this is a constructive and useful ethical stance to take towards your research and your practice, there can be some hurdles to cross in order to do so, including ensuring that all the appropriate people are consulted when proposing research and that clients who are potential participants are fully aware of the implications of taking part or withdrawing from the research.

Depending on your organization, the research topic and your methods of recruitment and data collection, you may have to seek approval for your research from any or all of the people involved, including your line manager, senior management, research supervisors, gatekeepers to participants, research and development committees and research ethics committees (Fox, Martin & Green, 2007). As a broad rule of thumb, the larger the organization, the longer it can take to meet all the required steps to gain ethical approval for your research, so factoring in time for this is an important consideration in your research planning. Remember too that ethics boards can request signatures and CVs from other researchers involved in your study as well as amendments to the original application, so time to ask for these also needs to be factored into the planning.

The importance placed on ensuring ethical research practice has seen the development of national and international frameworks to guide researchers.

The European Union for Cybersecurity Strategies is developing a Europe-wide governance framework for all EU member countries, and the Organisation for Economic Co-operation and Development (OECD) is working with the governments of 37 democracies with market-based economies to develop governance frameworks for the improvement and enhanced use of national health information systems in order to make better use of data for quality, safety and performance gains and to advance medical treatments and practices (https://www.oecd.org/health/health-systems/Health-Data-Governance-Policy-Brief.pdf, 2015).

In the UK, the Research Governance Framework was replaced by the Health Research Council as the UK Policy Framework for Health and Social Care Research (2017). This makes explicit the need to 'protect and promote the interests of patients, service users and the public in health and social care research by describing ethical conduct and proportionate, assurance-based management of health and social care research to support and facilitate high-quality research in the UK that has the confidence of patients, service users and the public' (www.hra.nhs.uk). These as well as governance frameworks within your organization provide essential and valuable reference points to ensure that your research is ethical.

Despite these widely available and comprehensive guides to conducting ethical research, however, there are some additional challenges particular to practice-based researchers. Perhaps chief amongst these is ethical considerations when seeking to recruit participants to whom you may be providing a service or product. It is always important to ensure that participants are provided with enough information to ensure they can provide fully informed consent to taking part. For the practice-researcher, this means going beyond the description of what is being asked of potential participants to make clear any implications for services or products they receive if they take part in the research. It is important to remember the potential for perceived power held by you if clients are considering whether to be a research participant. They need to be fully assured either that there will be no consequences impacting their services or products provision if they do not wish to participate or that they will be informed of the consequences of taking part.

As the researcher, you need to consider how you will ensure that you maintain boundaries between researcher and provider when working as a researcher with clients. You should ask yourself questions about how you may have these boundaries challenged during the research (perhaps by your client evoking a desire in you to help them during research interviews as they describe distressing experiences) or how you will manage information given to you as part of the research that you were not aware of as their service provider (perhaps describing experiences you had not previously been made aware of). It is important to think carefully and reflexively about how you will keep your researcher-self separate from your practitioner-self during the research – what support will you draw on to ensure this? Who can you debrief to after challenging data collection sessions? Many practitioners in health and social care settings can employ their skills in communication, rapport-building and attentive listening to conduct rich meaningful interviews with participants, but it is not appropriate to use interpretations about people that are not evidenced by data that you have collected. Therapists, counsellors and other practitioners make interpretations in the here and now as part of the service they deliver – this is very different from taking away verbal accounts provided by participants for interpretation through data analysis at a later date. If your research is to advance product development or testing, you have a responsibility to ensure that the research poses no risk to clients due to any defects in design, production or misinformation.

All research should ensure privacy and respect for participant data, but this anonymity can be challenging when recruiting participants from your own organization because it may be easier for them to be recognized by other members of the organization. Disguising personal details wherever possible helps to minimize this, and whilst sociodemographic data is important to some studies, it is not always the case, so think carefully about the value of including demographics such as age, sex, ethnicity and so on for public dissemination.

Observational methods of data collection can also present challenges for practice-based researchers. Overt observation may change the behaviour of colleagues or clients whilst covert observation can mean deceiving them in ways they may resent when they find out.

It is important too to consider existing power relations and perceptions of power relations when doing research with colleagues and clients. Think, for example, how difficult it may be for potential participants to say no to taking part if you are their line manager. It is your responsibility as a practice-based researcher to ensure, as far as you can, that all possible steps have been taken to respect potential client-participants and your involvement with them before engaging them in your research. If you do not feel able or comfortable doing this, then it may be better to conduct your research outside your organization.

Many of these issues may seem particularly pertinent to qualitative research but the Research Governance Frameworks make clear that ethical research considers the implications of all data collected from human participants. This takes it beyond interviews to include physical data, such as tissue or blood samples, which you may be collecting for statistical or performance analysis. Thinking through, and consulting others, about possible implications for participants and yourself will be beneficial to the quality and use of the research as well as to your status as a practice-based researcher, and may enhance the likelihood of conducting further studies.

In the next section, we consider a positive ethical stance and how this can further benefit your research and its status.

Positive ethics

Adopting a positive ethical stance in research means actively linking your personal values with the approach you are taking to the research. Positive ethics go beyond the 'floor approach' (Knapp, Gottlieb & Handelsman, 2018), which focuses on ethics rules and regulations, to consider what else you could do rather than simply taking a tick box approach to what you *should* do to avoid sanction by ethics approval boards. Positive ethics encourages researcher reflection on how to improve one's research. Combining a floor approach with a positive ethical stance not only better serves participants and researchers but also enhances the research standards.

In practice, positive ethical researchers seek to promote understanding and appreciation of individuals and groups, including those who are traditionally marginalized so that issues around discrimination can be addressed (Knapp, Van de Creek & Fingerhut, 2017). This means actively striving to maximize participant involvement in the development of the research by regarding participants as moral agents with intrinsic worth and helpful perspectives rather than as a means to achieving research goals (Fisher, 2000, cited in Knapp & VandeCreek, 2006). The focus of the research then includes power and its consequences, considering it from the outset and disseminating findings about effects of power to the wider world. This approach extends too to conducting research with colleagues and service users, so the positive ethical researcher seeks to enhance the quality of relationships with them as well. They take responsibility to update themselves regularly with relevant new knowledge and skills as an ethical duty that enables them to contribute to a range of communities (Frost, 2021). Key to achieving a positive ethical stance is striving to enhance trust throughout the research process, often meaning that confidentiality and consent issues go beyond one-off contractual transactions at the start of the study and, instead, become an ongoing and integral part of ethical reviewing of the study development.

Writing up and disseminating practice-based research

As is discussed in other chapters of this book, the writing up of your research is key. Without this being done, the research is, at worst, lost or, at best, confined to a very small audience. The writing-up process itself is a useful one for clarifying your thoughts, and research findings and outcomes, and, of course, in order for it to be taken seriously, the writing style must be clear and coherent with as much detail as appropriate to enable the audience to follow and understand the rationale, process and outcomes of the research.

For practice-based researchers, there are questions about how to best disseminate their research and where. Whilst it is clear for academic researchers that publication in academic journals is useful and necessary if they wish to contribute to the scholarly community, this is not always the case, or at least not the only case, for practice-based researchers. As we have said throughout this chapter, a key aim of your research is to motivate and bring about change in practice. It may be more useful, therefore, to publish research reports than academic articles so that they will be read by policy and strategy developers. It may be that verbal presentations of the research are more accessible for some audiences, or that research is disseminated through social media or online videos and blogs in order to reach the public (although see below for more on the risks of using these strategies). In this section, we will not only consider where practice-based researchers may want to disseminate their research but also what this means for how they present it.

To begin with, it is important to think about the objectives of your research; who do you want it to reach and for what purpose? Your primary research aim may have been to bring about changes in practice, in which case your key audience may be colleagues or other practitioners in your field. Or it may have been to provide solutions to key questions for industry partners. You may have carried out the research to inform policy development and, so, you want government policy makers to utilize the research. An important aim of your research may have been to increase understanding and awareness of a little-known, or misunderstood, societal issue, when your primary target audience was the general public. Being clear on why the research is being carried out and with what ultimate purpose and being able to communicate this to others will help plan for dissemination.

Gone are the days when academic journals were the only places for researchers to publish their research. Whilst these remain useful, even essential, options if you are researching with academics and want to reach audiences interested in scholarly work, it may be that professional, rather than exclusively academic, journals are better able to target fellow professionals.

Professional journals

Professional journals are scholarly journals targeting specific professional readers. Examples include *The Nursing Times*, *Variety* and *Writing Magazine*. Professional journals often differ from academic journals in appearance, with more visuals and shorter articles. Articles are not always peer reviewed and instead include pieces written by experts and professionals in particular fields. They do not have long lists of references but focus instead on disseminating up-to-date information about services, new trends and products. Authors have experience in the field and jargon is usually minimized. Often, these journals are available to individuals via subscriptions or through membership of professional regulatory bodies. They can be an excellent way of ensuring that those most interested in your work are being made aware of it.

Writing successfully for a professional journal not only means writing in a succinct and accessible way but also in ways that highlight the benefits, challenges and implications of your research. As with all published pieces, you want to attract attention, so a title that is meaningful is important. Rather than an abstract, your article should instead include a summary sentence or two under the title or at the start of the article so that it is clear what it is about and why people may benefit from reading it. You need to ensure that key messages about why and how your research is important to the profession include enough detail to be persuasive of the research's value. For some audiences, it may be important not to include too much technical detail, because it could obscure the implications of the research and risk losing readers. Including your contact details and inviting readers to raise queries with you or request more information from you enable further details to be shared. It can also be useful to let other professionals in your network know about its publication and invite them to submit a letter about it to the journal to draw attention and clarify your research's value.

The stable of professional journals includes magazines aimed at the general public. These may have more generalized titles targeting lifestyles, different stages of life or hobbies. If your research outcomes fit into this (perhaps on topics of motherhood or retirement, for example), it can be a good arena in which to disseminate aspects of general interest arising from your research.

Writing for popular magazines necessitates writing in an accessible language for the lay person, perhaps not outlining the research itself in more than a few sentences but highlighting its meaning and implications. It may be that your authorship is not included but it is also not uncommon for you to be 'name-checked' in the article as the researcher behind it. Depending on the magazine, an editor may ask you to be interviewed, and whilst this does not mean you should go into inaccessible detail about your research, it does draw attention to you and your research, which some readers may follow up on.

Other ways of disseminating your research include in-house newsletters in your organization, regular updates for funders and press releases. These will be shorter than journal publications, often giving only an outline of key aspects of the research and its outcomes but, again, written in a way that is accessible and invites feedback and contact.

Online dissemination

The rise of online technology has opened up myriad ways of disseminating research. You can set up a website or blog to keep interested audiences informed about the research you are doing. Some sites such as WordPress are free to set up, or you can prevail on your organization to set up a research-focused website. The content can be managed by you, as can the levels of interaction with readers, and such websites can be a good way to form new collaborations.

However, a word of warning, these routes mean that you must strive to keep the postings up to date – readers will quickly tire of out-of-date information and/or assume you are no longer researching in this area.

Social media offers fast and wide-reaching ways to disseminate your research. Your research may be further disseminated by initial readers, and you can attract interested audiences with links to fuller articles and reports. There is also. of course, the advantage that other users of social media can quickly and efficiently respond to you. This method of dissemination has been shown to provide measures of who the research is reaching and how people are responding to it. Ways of including mentions on social media are developing to assess research impact by looking at who is downloading it, how people are reacting and the levels of engagement with it. This can inform you, for example, if the groups you want your research to reach are accessing it. Altmetric.com enables you to capture and collate online conversations about your research so that you can monitor and report on it for assessors and funders. The European Commission policy for the dissemination and exploitation of research results (https://research-and-innovation.ec.europa.eu/strategy/dissemination-and-exploitation-research-results_en) aims to help make results available to the 'people that can best make use of them' and to build support for future research and innovation funding, ensuring the uptake of results and opening up of potential business opportunities. By focusing on research results, the policy seeks to translate research concepts into concrete solutions that have a positive impact on the public's quality of life.

Ways of using social media and the range of technologies are rapidly increasing, and there is no doubt that they offer dissemination of research at a pace that is faster than traditional journal publication. However, there are some important potential pitfalls to be aware of.

The snappier online presentation of your research makes it more widely accessible, but it also means that some readers will not be interested in pursuing it further to acquaint themselves with its scientific rigour. Publishing your research on social media means, of course, that you cannot control what it done with it. This risks misinterpretation of your results being shared, used or presented incorrectly. It is important, therefore, to find the balance between presenting information that is interesting to the public whilst also maintaining accuracy in what parts of your research you report and how you report them (Smoliga & Kendall, 2017). There are calls to develop online review systems to attract input from experts about the validity and quality of research claims made on the internet (Williamson, 2016). It is important too to ensure that the reporting of your research online is ethical and does not intentionally bring false hope or beliefs to readers who may be looking for solutions to personal health or social problems (Dijkstra, Kok, Ledford, Sandalova & Stevelink, 2018).

All that said, however, the possibilities of using social media for the dissemination of your research are increasing all the time, so with careful attention to what you say about your research, how you say it and who might be reading it for what purpose, it can be an effective way of reaching others, including those who may want to network with you or offer potential funding for further research.

So far, we have focused on written forms of research presentation. It is now time to consider other ways, including verbal and visual forms.

Verbal and visual dissemination

Conference presentations are a long-standing way of disseminating your research in environments that allow you to interact with audiences, consider new queries about it and receive constructive feedback. Now that the importance of research in practitioner contexts is widely accepted, there are many opportunities to do this. A conference can not only be a productive way of sharing your research but also a way of learning about other research being carried out in your field.

The format of research submissions has extended in recent years; so, now, in addition to the traditional paper and poster presentations, prospective presenters can choose from a range of dissemination options.

If you are more comfortable speaking directly with others about your research than 'presenting' it to them, you can consider being interviewed about it in front of attendees. You can agree on key points or questions with your interviewer (usually someone who knows your research or is in the same practice as you) in advance and rely on them to encourage audience interaction, either at the end of the interview or as points are made during it. If you are comfortable, you can present your research and act as your own facilitator, asking questions of and answering questions from your audience (sometimes known as 'fireside chats').

World cafes are organized around several tables of six to eight people, where you discuss a key issue or question related to your profession or industry. After a specified time (typically 15–20 minutes), each person moves to a different table to use these discussions as the basis for a second, related issue.

Fishbowl sessions allow for three or four people to publicly discuss their research or an issue, sometimes being joined by an audience member who may swap out with another during the discussion.

Faster-paced sessions include Pecha Kucha, in which presenters spend 20 seconds presenting 20 images to highlight key points of their research. Whilst there is not enough time to discuss the research with the audience, this approach draws attention to it and can always be followed by a discussion with interested conference delegates over a cup of coffee. Fast three-minute presentations are also increasingly common at conferences and can be a useful way to think about how best to consolidate the presentation of your research.

Whilst words, written or oral, remain a common way of disseminating research, it can also be done through online videos and by attracting audiences through visual abstracts. Visual abstracts use images to convey the overall ideas of the research, summarizing it and aiming to draw interest to it. They can be developed with diagrammatic or infographic approaches but the key is to ensure that the key points are conveyed well in a self-explanatory way. These can be particularly effective in reaching those who are interested in your research but may not have the time or inclination to dig out full research papers, because quick decisions can be made about their interest (and, therefore, follow-ups) on the basis of the visual abstracts. This can result in increasing interest in your research from people who really want to know about it, whether they be fellow researchers, service users or policy makers. Visual abstrats can be used on social media effectively, but are also being introduced into some professional journals (Ibrahim, Lillemoe, Klingensmith & Dimick, 2017).

Finding time

It is hoped that the chapter to this point has been upbeat and positive about how to be a practice-based researcher and have your research recognized, but it is well known that for many practitioners, despite their desire to research, the primary barrier to being able to do so is a lack of time (Valentino & Juanico, 2020). Some of the reasons for this have been included in the discussions above, including poor recognition of the value of research, lack of support for doing research and high workloads. Despite these and other potential barriers, there are ways to have time to research as a practitioner and we shall discuss some of these in this section.

Make time

Perhaps the most useful tip is to *make* time rather than *wait* for time to do your research. This can mean writing out (and sticking to) a schedule for research, allocating particular hours on particular days and creating 'new' time (early in the morning or late at night, for example). Treating these allocations of your time as work commitments that cannot be broken is likely to ensure that the time is there for you to do your research.

Planning the time

In addition to making the time, it is important to plan what you will do with it. Knowing the likely schedule for the various aspects of your research will bind you to what you do with your allotted time and means that you will be more likely to meet research deadlines.

Using time

Many practice-based researchers lament the lack of days free from practice obligations when it comes to developing and carrying out research. There is a belief that unless you have large blocks of time, you will not be able to do the research. This is not the case: even a couple of hours can be used for research admin, recruitment, data analysis and so on. Steady progress is achieved, which can be rewarding and motivating in itself. A useful tip when it comes to writing up research is to leave notes to yourself about what you will write about when you return to your writing. This takes away the time taken in thinking about where and how to start your writing session.

As a practice-based researcher, you have the advantage that it can be possible to incorporate research into your practice. Instead of seeing the two activities as separate and exclusive, some professional practices (such as teaching, for example) can be used for observation or evaluation (always ensuring ethics have been considered and approved), thus negating extra time needed for participant recruitment (Valentino & Junaico, 2020).

Challenging time

Finding time to research can be part of a mindset. Instead of telling yourself you are too busy with your practice to carry out research, remind yourself of your enjoyment of research and the value you place on it to evaluate how you use the time you have. Are there periods when research can be prioritized over practice? Or times when doing research will enhance your practice?

Review your understanding of how research in practice can be done. It won't be like previous studies or research roles, where the expectations are that you devote long periods of time (months or years) to doing research. Instead, you can see research and its outcomes as part of your work responsibilities, benefitting you, your practice and the profession more widely.

Accepting time

Valentino and Junaico (2020) advocate patience for practice-based researchers. External events and practice pressures which delay the research may occur, or services and products may develop in your research direction faster than you expected. That's okay – you are not on a deadline, as you may have been during your training, and instead are looking to contribute meaningfully to a profession. If necessary, change the participant group, the research question or indeed the research topic. With commitment and a passion for being a research contributor, you will do research that you are proud of and see as making an impact.

> **Reflective question**
>
> How do you find or make time for research? Does it help or hinder your professional practice?

Chapter summary

In this chapter, we have focused on the benefits and challenges of being a practice-based researcher. We have considered some of the practicalities of conducting and disseminating research in your workplace, among your clients and in the field more widely. We have looked at ways in which collaborations can be formed to help you to be part of and complete research, and we have discussed ways of ensuring your research has its potential for impact maximized. We ended the chapter with a focus on how to make, manage and organize time for research when you are also a practitioner.

In Chapter 6 we focus on the role and work of the community researcher.

Further reading

Bager-Charleson, S. (2014). *Doing practice-based research in therapy: A reflexive approach*. Sage Publications.

This book makes the vital link between practical research skills and self-awareness, critical reflection and personal development in practice-based research. It guides the practice-based researchers step-by-step through the practicalities of the research process, encouraging them to reflect upon and evaluate their practice at each stage.

Schwabish, J. (2016). *Better presentations: A guide for scholars, researchers, and wonks*. Columbia University Press.

This book for presenters of scholarly or data-intensive content provides essential strategies for developing clear, sophisticated and visually captivating presentations, using three core principles of visualize, unify and focus. It includes practical tips, clear examples for what to do (and what not to do) and shares the best techniques to display work and win over audiences.

Wilkinson, D., & Dokter, D. (2023). *The researcher's toolkit: The complete guide to practitioner research*. Taylor & Francis.

This book provides a practical and accessible guide for practitioners undertaking small-scale research for the first time. It covers the entire research process, from defining a research topic or question through to its completion.

6 Being a Researcher in a Community Setting

Introduction

In this chapter, we explore in-depth what it is like to be a researcher conducting research in a community. Often community-based researchers are members of the community and this brings particular strengths and challenges to their research practice. We shall explore these and how to manage them. We start the chapter by considering what it means to be both a member of a community and a researcher within it, and we move on to discuss how community members become community researchers. We then consider the challenges and benefits of researching your own community. We consider how to explain and justify the research to other members of the community and ways of managing unwanted research or research outcomes. We focus on ethical considerations particular to this type of research and how to negotiate challenges and tensions that can arise between you and your fellow community members. We finish with a detailed look at effective ways of including community members in your research and keeping them involved from research design to output.

Being a community-based researcher

Being a community-based researcher means first understanding and defining what you are considering as the 'community'. It is only then that you can reflect on your membership of it and what perspectives you can bring to carrying out research with its members. A community is a group of people who have a characteristic (or characteristics), or shared interests or attitudes in common, or who live in the same place. What makes you a member of the community with whom you are doing research, and what understanding (or experience) of the research topic will you bring as a researcher? This is important to think about because a crucial aspect of being a community-based researcher is to develop – in the community members – trust in the research. This trust will be in you as a researcher as well as in what will be done with, and as a result of, the research being carried out.

How this trust can be achieved needs to be thought about from the outset. Knowing what makes you a member of the community (and what does not) can help build rapport and minimize assumptions in the relationships with community members. If your understanding of the community and your membership of it are incorrect or misinformed, trust may be broken and data may be misinterpreted.

A primary aim of community researchers is to help bridge the gap between the community and external institutions. You may be carrying out the research with academic co-researchers and other stakeholders based outside the community and it will be important to your role to bring a nuanced understanding of the experiences of community members, to enhance access to community members and to ensure community involvement in the design and conduct of the research. As a member of the community, you may be better placed to convey messages about its purpose and potential outcomes to other members, in the hope that you will cultivate interest and conviction in the research purpose. Community-based researchers will have a better understanding of social and other structures of the community than those outside it. You may have personal experience of the research focus, so your insight into it may have more subtle and valuable shades of meaning than that of researchers outside the community, and you are more likely to have a better understanding of why communities may feel they have been let down by previous research studies. As a community-based researcher, you want to bring about impactful change through the research by bringing a perspective that enables greater comprehension of how the research issue is understood within the community.

It is important to understand, however, that simply being a member of a community does not mean that as a community-based researcher, you are an insider in all other aspects of what it means to be a member of that community; in addition to some of the commonalities of the community member perspectives, you will have differences. These may arise from your experiences in the community and how you have made sense of them, based on your gender, age, sexuality and other dimensions that make up your identity and worldview. There are also many ethical considerations to be thought about when you are researching with people from your own community – issues of confidentiality, information-sharing and anonymity, to name a few. You may also have to deal with research results that are unwanted by the community, and this needs to be something you prepare for. In the next sections, we will consider these issues and how to address them.

Being an insider/outsider researcher

Community-based researchers are often referred to as insider researchers – researchers who share some of the cultural and other characteristics of the group they are a member of (Loxley & Seery, 2008) and who have *a priori* intimate knowledge of the community and its members (Hellawell, 2006). However, the reality is that you may be an insider on some dimensions and experiences, and an outsider on others. Being a member of the community you are researching may well mean that you (and others) assume you will understand nuances of language, humour and cultural references made by fellow community members. Similarly, sharing the experience you are researching with other members may mean you believe you have common understandings of its meaning and impact on the community. It would be a mistake, however, to assume that having *a priori* knowledge of the community implies the meanings you make of other community members' accounts are the meanings intended by them. To reduce the likelihood of your pre-existing knowledge of the community being imposed onto the data, it can be useful to think about your positionality in the research process.

Researcher positionality

Researcher positionality recognizes that enduring social identities can confer a status that enables or limits the exercise of power. It is an important concept central to rigorous critical research. Different identities come to the fore in different contexts and social interactions, as they intersect with constructs of identity such as race, social class and gender (Frost & Holt, 2014). As a consequence, different positions are taken up in accordance with your worldview, your position in relation to the research task, and in the social and political context in which the research is being conducted (Holmes, 2020). By considering positionality in insider research, Chavez (2008) distinguishes between *total* insiders (sharing multiple identities or profound experiences with the community) and *partial* insiders (sharing a sole identity but with some distance or detachment from the community). In practice, researchers take up varying positions in the research process, which are at times more and less salient to their insider status.

Being aware of how you position yourself during the research process and of how you may be positioned by others can enhance your reflexive awareness of why participants say some things, and not others, and of assumptions you may bring to the interpretation of their accounts. Being a community-based researcher can mean you have to work harder to develop and maintain your reflexive engagement with the research so that you can understand the influence on your positionality of the assumptions you bring (see, for example, Ganga & Scott, 2006). Other community members may perceive aspects of you as a community member that you had not been aware of – think about your accent, for example, or periods of absence that you may have had from living in the community. Participants may assume that these lead you to have different understandings of what it means to be a member of the community and may choose their responses in data collection accordingly, for fear of being misunderstood, criticized or judged. It may also be that they wish to go beyond the short answers required by surveys or questionnaires to contextualize or share aspects of their experience that you have not enquired into (see e.g. Ryan & Golden, 2006). It may also be that if unforeseen fissures and divisions in the community become apparent in the research, you may be positioned by some community members as belonging to a subgroup within them.

The key message to remember when working as a community researcher, therefore, is to be as attentive as you can be to how you see yourself as belonging to the community and to be aware that this may differ, on some dimensions and at different contexts, to how other community members see you.

The following reflection, taken from Kerstetter's (2012) work on the impact of researcher's identities on the community-based research process, illustrates how insider researchers can also come to feel like outsider researchers.

Researcher reflection (from Kerstetter, 2012)

I was an insider as a Maori mother and an advocate of the language revitalization movement, and I shared in the activities of fund raising and organizing. Through my different tribal relationships I had close links to some of the mothers and to the woman who was the main organizer...When I began the discussions and negotiations over my research, however, I became much more aware of the things which made me an outsider. I was attending university as a graduate student; I had worked for several years as a teacher and had a professional income; I had a husband; and we owned a car which was second-hand but actually registered. As I became more involved in the project...these differences became much more marked. (pp. 137–8)

Reflective question

What communities do you consider yourself to be a part of? What is it about you, your experience and your worldview that leads you to think this?

Ethical considerations

The ethics of research conducted as a community-based researcher require additional considerations to ethics in other research; of course you should adhere to the expectations of minimizing distress brought about by participation in the research and respect standard criteria such as obtaining informed consent and debriefing participants after they have taken part in the research. Ensuring these in practice can sometimes be more challenging if you are also a member of the community. As a community-based researcher, you are also likely to uphold positive ethical stances (see Chapter 5) because of your desire to bring about beneficial change for the community. It is also important to think about context-specific ethics for the community in which you are conducting research, and we will focus here on what that can mean in practice.

Most universal ethical principles to which researchers are expected to adhere have a Western individualized focus (Mavhandu-Mudzusi, 2023). This means that issues of individual respect, privacy and confidentiality are emphasized. Whilst important and applicable in some research contexts, these issues remind us to consider how the community we are conducting research in operates. For some, the individual is seen from the perspective of significant others in the community. For example, Mavhandu-Mudzusi highlights that in rural areas of South Africa, some communities have a higher degree of interdependence between a community's members. Elders may expect to act as gatekeepers and women may be expected to have chaperones, usually other women, present during interviews or other data collection. Rather than seeing this as challenging anonymity, it can be seen as an appropriate and respectful way to conduct the research.

Van Zyl and Sabiescu (2020) extend this argument to propose intersubjective ethics when conducting research with vulnerable, minority or Indigenous communities. Rather than relying on prepackaged and prepared ethical protocols and forms issued by the research institution, these researchers advocate emerging, situated and negotiated ethics in research. This means recognizing that differences in paradigms and worldviews may exist between researchers and communities, rendering the research adhering to traditional (in the West) ethics practices potentially impossible, misleading or misunderstood. Intersubjective ethics calls for a grounding of ethics approaches in intersubjectivity so that shared understandings and practices are developed and maintained. Transparency and agreement are called for in issues of epistemological and ontological assumptions brought to the research, in the vocabularies used and in the mutually agreed and shared ethics protocols.

In practice, a key way to raise your awareness of how your ethical practices constrain and are constrained is to strive towards *being* in the field. Recognizing your lived experience and what it means in terms of being a researcher helps to enhance an unobtrusive presence, empathy, patience and suspension of judgement. Having genuine curiosity about the community's history and the history of the topic under research brings an openness to understanding socio-cultural beliefs and practices, and offers you opportunities to question your own experience, assumptions and biases about them. Considering these in how you make transparent, develop and practise ethics in the research can enhance trustworthiness and co-research with community members, as well as enhance your openness to receiving knowledge from them about how to design and conduct the research. Furthermore, being reflexive throughout the research process will better position you to recognize ethically important moments when they arise and find ways to address and manage them. Whilst many of these aspects of reflexive, intersubjective ethical practice can be learned and shared, it is important to remember that as you become more attuned to your lived experience of being a community-based researcher, you are likely to increasingly internalize and develop them.

Enhancing recruitment and partnership

A key role of community-based researchers is to attract other members of the community to the research as co-researchers and participants. Often, you have been recruited as a researcher because of your community membership in the hope that this will give you greater access to other community members and a valuable understanding of the relevance to and implications for the community of the issue under research. It is important, therefore, to be prepared to explain and, if necessary, justify the research to fellow community members.

Some may be mistrustful of any benefits coming from the research and others may be wary of what the research might mean for their lifestyles and practices. Some may simply be tired of taking part in research that rarely seems to result in change.

As a community researcher, you are important to the recruitment process. When explaining the research to potential participants, gatekeepers, and co-researchers, you need to consider not only the language you use and how you present the project but also why community members should invest their time and energy in it. The relevance of the project to both individuals and the community as a whole has been found to be a key consideration for community members wondering about participation (Pelletier, Pousette, Ward & Fox, 2020). Community members need to have reasons to engage with the project, so the direct value and benefits of the research to the community and its individuals need to be conveyed in ways that can enable potential participants or co-researchers to see the point in taking part in it. They need to know how the research outcomes will be used and whether they will be impacted by them. Of course, the most straightforward way of finding out what important issues are most likely to attract people to the research is to ask them. As a community-based researcher, you will have some relationships and networks that you can consult and by involving community members in the crafting of the research question, you can take a further step in ensuring the research is of relevance to them.

Communication has also been found to be important in attracting and retaining community members in research projects. This goes beyond using accessible language to include regular two-way information sharing on the progress (or lack thereof) of the research and what is happening both in the process and with the outcomes. This means an investment of time for community researchers that will keep them involved with the community after the data has been collected – updates on analysis, report writing and action based on the research should be offered regularly, alongside an open-door policy for community members to maintain contact with team members and managers. Pelletier et al. (2020) also found that members of some smaller, rural or remote communities sought reassurance that the information they provide will not be obscured or suppressed by data from larger communities. Preferred communication mechanisms should be identified so that you are clear about how best to connect with community members. Sometimes, this will be through elders, research champions and councils and sometimes, through verbal presentations, email or social media. It can also be useful to find out about local news outlets such as radio and newspapers, particularly if communities are less connected with national or technological forms of communication.

Taking these approaches to describing the research and explaining its aims and purpose will help develop meaningful projects which community members will want to be involved in. Regarding community members as experts on the research issue will enhance trust and is more likely to lead to meaningful research interactions and co-generated knowledge. By promoting and practising ways of working in partnership, concerns about power dynamics and the value of the research may be addressed, which means that the research outcomes are more likely to be useful and used. Striving for this starts at the beginning of the research process when you are explaining and justifying it to the people you want to involve in it.

It should be said, however, that researching with your own community, even if done well, does not always bring the expected or wanted research outcomes. In the next section, we will consider what this can mean to a community-based researcher and how it can be managed.

Managing unwanted research outcomes

However successful you may have been in developing a partnership with community members in order to carry out research with them, it can still be the case that the outcomes of the research are not what was wanted. Think of evaluations of community services or of health research that aim to change behaviour, for example. There may be a wariness of community research leading to perceived interference by outsiders or to an imposition of new policies and practices that may be in tension with those that are long-established. Developing meaningful partnerships can help to alleviate the management of the delivery of unwanted research outcomes. Maintaining regular and frequent channels of communication helps prepare community members for where the research seems to be going and ensures that the issue being researched continues to be relevant to the community.

It is also the case, however, that research designed to bring about change risks bringing about unwanted change or causes lives to be impacted in ways that are not wanted.

Although never easy, the tension arising from sharing unwanted research outcomes can be mitigated to some degree without compromising messages about the value of your research. Openness from the outset that the research may lead to unexpected or unwanted findings helps prepare those who may be affected. Highlighting from the start that negative results are possible and being prepared to listen to questions and concerns about what these might be and what they might mean for the community help to maintain trust. Bearing in mind what we have discussed above about the relevance of research to individuals as well as to the community being an important factor in encouraging people to be involved, it will be sensible to tailor your explanations to convey how such results can impact individuals as well as the community as a whole. Emphasize (and ensure) that the research team is keen to consult the community members and discuss the research with them and will talk with them about decisions that have to be made as the research progresses. This is part of a partnership approach and will serve to aid understanding in how the unwanted results have been reached and what their implications are. This might include being clear about who has commissioned the research and why, and who the research outcomes will be disseminated to and for what purpose. Similarly, try to ensure that the results are heard directly from you or the representative of the project with whom most communication has been held. This respects the trust that should have been built during the conduct of the research, and bad news is likely to be better received this way than if heard second hand or inaccurately.

It can also be useful to allow time and interaction with the community to plan and implement changes in response to findings, as problems are identified during the research. Although not all will be immediately identifiable, or implementable, the fact that the need for change has been recognized and plans to address them made will show the value of the research being carried out even before it has been completed. Related to this, you can make and gather suggestions for how to address some of the unwanted results and provide contacts for people with whom follow-up discussions can be had. It is also useful to show how other results from the study have brought successful change and how this too can be built on. Sometimes, the positive findings and implications of the research may be tangential to the main research focus but have a bearing on how change can be implemented. Almost always, they work to show what is working well or will be responsive to change that will benefit the community.

The researcher after the research

When the research ends, community-based researchers will still be part of the community. It can mean that former relationships with other community members have changed and that new ones have formed. As a community researcher, you may have information about fellow community members that you would not otherwise have had or they may have learned about aspects of you that they otherwise would not have done. It can also be that community members are disappointed in the research outcomes or in you as a researcher. The problems related to navigating relationships with former participants can be mitigated to some degree by being clear and transparent from the outset about the purpose and design of the research as well as about your role in it. Reviewing and discussing this regularly can encourage participants to raise questions and concerns as they arise. Striving to maintain your researcher boundaries, and recognizing when you do not, can minimize and manage self-disclosure that you may later regret, and, of course, adhering to ethical requirements of confidentiality is paramount. Being open with community members about why the research is being conducted and at whose request helps manage expectations of it. Explaining what it may achieve, but also that there may be a range of outcomes, helps with trustworthiness. As with all research, potential participants should be fully informed about what is required of their participation and how the outcomes will be disseminated, and this too should help reduce participants' surprise or annoyance.

Community research is usually conducted with the aim of bringing about change. This means that the research focus, why change is being sought and what changes might result should be clearly explained. Remember that change is not always welcomed and that there may be fears about how it will impact lifestyles and behaviours. Community members may be wary of increased or enforced regulation as a result of the research and this too can cause resistance to the research. If you are open to questions and provide clear explanations of the research, what it entails and what it might lead to, you can be more confident that you are bringing at least some of the community members along with you, both as participants and co-researchers.

Training to be a community-based researcher

Training to become a community-based researcher is likely to be offered if you are not already working as a researcher in other settings. It may be delivered by academic institutions with joint interests in the research or by community-based institutions who commission research with community members. The trainers will first aim to recruit potential community researchers by providing knowledge of the research focus, how and why the study will be carried out and what the research will be used for. The recruitment will also emphasize the personal characteristics that successful community researchers have and what will be expected of them in the researcher role. It should include a question-and-answer session so that you can raise any queries.

If you confirm your interest at the end of the session, you will be included in a pool of potential researchers, some of whom will form a core group and others who will be kept in reserve in case of attrition during the research. Whichever group you are in, you can expect to receive full training.

To be selected for training, you will need to show an understanding of the research project and the role of the community-based researcher and that you can work well collaboratively, with good listening skills and confidence in talking to others. Recruiters will want to know that you have time available and the commitment to take up the role and that you understand and will practise the ethical standards of the research. The recruiters may assess for these values in one-to-one interviews and/or group exercises.

The length of time taken to provide the training will differ with every institution, and you can usually expect to have ongoing mentorship and supervision when it is finished. The training is likely to use a variety of information provision sessions, knowledge sharing about how to do research, group activities for team building and research skills development, and role play. The training will aim not only to equip you with the knowledge and practical skills of being a community-based researcher but also with the underpinning principles of learning about communities and how research can bring about change. There will be an emphasis on who you are as a person, so you may be encouraged to share personal experiences, characteristics and why you are interested in becoming a community-based researcher. This may feel uncomfortable until you have built trust with the trainers and other co-researchers.

By the end of the training, you should feel that you know enough about the project and what is expected of you in it to enable you to make an informed decision about whether you want to take up the role. You should feel confident that the research has been well-planned and will be monitored throughout, and, also, that you will have support when you need it as you conduct the research. You should feel that you have been trained well in how to elicit and gather meaningful data in the forms required by the project (interviews, surveys, observation and so on). You should also be clear about the limits of your role and what to do if you encounter challenges in carrying it out – whether the challenges are from life events or the research process.

Some community researchers are recruited to gather data and pass it on to other researchers for analysis, whilst others are involved in the data collection and analysis stages. There is benefit in community-based researchers being involved in data analysis because they can bring their unique insight as a member of the community being researched to recognize nuances and meanings that researchers outside the community may not. To be involved meaningfully in data analysis usually works best if you are involved in the design of the project too so that you can fully contribute to how and what data is gathered.

Finally, you should have confidence in the management of the project. Has the training been sufficient? How can you give feedback during and at completion of the research? Is the timing of each phase realistic and feasible? Are you supported in maintaining relationships with the rest of the research team? All of these and many other aspects will filter through from strong, supportive and efficient supervision. It can be useful to chat with project managers prior to accepting an offer of a community-based researcher role so that you understand how they aim to work and how they will ensure the quality and value of the research.

With all these aspects in place, your role as a community researcher can be rewarding and empowering, carried out in safety and with support from others when challenges arise. This is not only fulfilling for you but should imbue a sense of the importance of the research in you that you, in turn, will transmit to others in the community with whom you will be conducting research.

Doing the research

This book is about being a researcher and not about how to do research, but in this section, we are going to outline the Participatory Action Research method because of its intense focus on engagement between researcher and participant and how community-based researchers uphold the principles of community research. Community-based research aims to be conducted with communities in collective, interactional and democratic ways, to bring about change. Critical to this is the reflexivity of the researcher who is required to consider throughout the research what it means for them to be both a member of the community and a researcher of it.

Participatory Action Research

The Participatory Action Research (PAR) approach increasingly underpins community research. It is an approach that '...prioritises the value of experiential knowledge for tackling problems caused by unequal and harmful social systems' (Cornish et al., 2023). It seeks to foster partnerships with communities and to involve community members in all aspects of the research design, conduct and dissemination. This requires relationships between researchers, co-researchers and community members, and this is why community-based researchers are so important to community research. In this section, we will take a closer look at their role in this approach.

Community-based researchers are often recruited because of their existing knowledge of the community, how it works and the problems it faces. This knowledge comes from being a member of that community. They have relationships with community members and not only know individuals within it but also have a greater understanding of who may be interested in and necessary to carrying out the research.

Community researchers are offered training and support to play an active role in the research (see previous section) and this can be helped through university-community partnerships. Involving academic institutions in research brings benefits such as administrative support, researcher training opportunities and sustainable staff appointments. It can enhance the status of the research and offers another strand of research dissemination through academic journals, highlighting and potentially extending the research and its value to other projects. University researchers may be contacted by community representatives to initiate research, or universities may be approached by communities because they are aware of problems within them that can benefit from research.

Community-based researchers can benefit from these partnerships but can also be challenged with protecting the relationships and networks they have with the community so that they are not used simply as a source of knowledge and resource extraction. The relationships between the community members and the community-based researchers are key to the success of the research and preserving and developing these form a community-based researcher's main priority. The community researchers can act as a conduit between academic researchers and the community, facilitating interaction between the two so that the communities, perhaps distrustful after having been previously treated as passive objects of research, can ask of the academic researchers 'Who are you? Why should we trust you? What is in it for our community?' (Cornish et al., 2023). The community researcher treads the line between their own relationships with community members and their relationships with the university and researchers involved.

Cornish et al. point out that PAR projects are often part of longer-term collaborations – seeing the research through to the implementation of the changes it is seeking to bring about is not always possible in one short project. Continuing work on or extending existing collaborative relationships can build over time to develop increasingly meaningful and applicable research. More than one short-term project can be carried out and temporary increases in numbers of researchers can be enabled as necessary. The relationships can strengthen and co-researchers can become increasingly empowered, as participation and co-research develop and trust builds.

For successful community-based researchers, the building of trust and partnership begins with identifying and agreeing on the research focus. Defining the research problem can be an outcome of the PAR as well as the focus of research carried out. It can mean assessing the capabilities, skills, priorities and resources of co-researchers so that feasible, actionable research questions that hold meaning for the community are developed with the team and the community members. Community meetings, workshops, drama and focus groups can be held to identify and home in on key community concerns for research. The activities and the discussions contribute different foci and forms of expression which can be used to craft detailed research questions. In these contexts, the research questions may be different to those asked from academic perspectives, often with a more practical focus. Communities want to see realistic pathways to change through the research and this can mean that questions are asked about steps to be taken, exploring the potential of desired outcomes or evaluating ways in which the problem is currently operating in the community.

The community-based researcher working with a PAR approach, therefore, needs a diverse range of skills and personal characteristics. As well as a willingness to learn about conducting research, they will need to be able to work as a team member, be creative in identifying research problems with others, have ways of communicating effectively, sensitively and respectfully, be open to acquiring knowledge and experience and be reflexive about their own role and work in the project.

PAR projects are often conducted iteratively with frequent review of the actions and their outcomes to allow for redefinition of the problem and actions taken. Community-based researchers, therefore, must have commitment to the research, an open mind as to how it is conducted and the capability to hold tensions and discomforts that can arise from the process. They need to be able to observe as well as to participate and to be willing to discuss with others what they see as going well and less well in the project. Given that successful PAR teams can involve stakeholders and co-researchers who each bring different perspectives to the project, the community researcher will need to have flexibility in how they understand the community and the world it inhabits.

A successful, large-scale PAR project was conducted by Veale, McKay, Worthen and Wessells (2013) to explore an intervention with young mothers formerly associated with armed groups in Sierra Leone, Liberia and northern Uganda. The project is outlined below to illustrate the design and effectiveness of the project.

Summary of a PAR research study, adapted from Veale, McKay, Worthen and Wessells (2013)

Background: The group of national and international university-based researchers (Veale, McKay, Worthen and Wessels) developed a team comprised of NGO agency personnel and community members drawn from three Sub-Saharan African countries. The project aimed to work with vulnerable young mothers, many of whom had been associated with armed forces or armed groups (children formerly associated with armed forces/groups (CAAFAG)), to support their social (re)integration into their communities through the process of taking up meaningful and socio-economically sustainable roles. The participants were active as decision-makers throughout the research and the group methodologies used in the project aimed to facilitate them to identify and implement social actions that they felt would best achieve this. The participants identified priorities that included better health, better social relationships and economic self-sufficiency (see McKay et al., 2011, for more on this) and the key principle of meaningful participation was upheld throughout to better ensure that the research and its priorities were always determined by the girls and women taking part in the project. The particular focus of the project was to explore how, through the PAR, young mothers engaged with traumatic stressors such as marginalization and disempowerment to take control of their lives for themselves and their children.

Research question: The project asked 'Did participation in the PAR facilitate young mothers' social (re)integration?' It addressed the question by enquiring into how psychological functions such as managing anger and overcoming shame and social relationships evolved and changed through participation in the activities of the group. Overall, the project was interested in whether young mothers' engagement as decision-makers in activities instigated by them would lead to improved integration and decreased stigma.

Method: Community members, NGO agency personnel and young mother participants in 20 study sites in rural and urban communities in Sierra Leone, Liberia and Uganda recruited 658 young mothers for the study. They had more than 1,200 children between them. The mothers had either been formerly associated with armed groups or armed forces or were considered particularly vulnerable. The consent form (translated into the Indigenous languages of each study site) emphasized that the study wished for each woman to be central to the planning, implementation and evaluation of the project to help researchers understand the information provided. In addition, Community Advisory Groups were established to help the mothers and agency staff plan social actions to address their priorities. Actions developed included drama, dance and poetry that served to challenge misconceptions and give insight into the community members' lives. Group savings schemes were started as well as group gardens, small trading, food businesses, goat rearing, community cleaning, home visits and help at funerals. Skills-building initiatives were introduced throughout the duration of the project. These included managing group dynamics, business skills, health and hygiene, and gathering and analyzing data. Annual meetings were held with academics, team leaders, representatives of the young mothers, donors and invited UNICEF experts from the three countries. Documentation included monthly reports from individual field sites, training workshops for young mothers, transcripts of team meetings, ethnographic reports and a participatory evaluation survey. In addition, regular field visits were conducted by the academics who met with young mothers, agency staff, local leaders and country team leaders.

Results: Key results included 'Moving from Other Regulation to Self-Regulation' and 'Shifts in Identity, Role and Community Membership'. The young mothers described how taking part in meetings over time helped recognition of common elements amongst them and that other members had their own challenges that meant they related differently to the community. In reaching key moments, when the groups faced choices about how the young mothers related to each other, they formed ways of learning to function as a group. As well as adopting modulated tones of voice and careful choice of words with each other, the young mothers began to share their personal stories, establishing a sense of solidarity. They began undertaking a variety of social actions which highlighted the value of the group and changed the self-efficacy and the self-concept of individual members. This shift from other-regulation to self-regulation enabled a taking over of responsibility for tasks and imbued a sense of ownership and agency amongst the mothers.

Summary: Examining how participation functioned as a principle and tool in community reintegration highlighted the promotion of individual and collective efficacy and social connectedness. It enhanced participant capacity to address priority survival needs and showed how communication styles and behaviour were adapted to improve personal and community relationships and inclusion into civilian society. The adoption of the participation principles by the PAR team and other stakeholders facilitated them to question and review with community members the ways that changes could be brought about, always keeping the priorities identified by the participants at the forefront of decision-making.

The focus on the project above demonstrates the potential value of PAR approaches in participant-led research. The extent to which the researchers were part of the process and whether they were community members or not, were shown to be key to the successful implementation of PAR principles. This impressive project was a large-scale, multi-site study which included stakeholders bringing many different interests and perspectives. Its focus on how and why PAR approaches can be respectful, empowering and effective gives many signposts for other projects, be they small or large-scale and in a wide range of communities. The mantra of the researchers in this project was 'If it doesn't come from the girls, it's not PAR' (p. 831) and provides a key tool for PAR research in all contexts.

Chapter summary

This chapter has discussed some of the expectations, challenges and personal characteristics of being a community-based researcher. It has highlighted the importance of understanding your position and role as a community-based researcher and of establishing trust with community members. By focusing on Participatory Action Research (PAR) as an approach that many community researchers used, it has shown the importance of finding ways of involving community members as participants and co-researchers so that this approach in itself can become a tool that is beneficial to communities.

Further reading

Bell, S., Aggleton, P., & Gibson, A. (Eds.). (2023). *Peer research in health and social development: International perspectives on participatory research*. Routledge.
  The book seeks to counter the marginalization of research experience and skills derived from close relationships with people and communities, while reflecting critically on the strengths and limitations of peer research. With a wide range of international contributors, it illustrates the potential of peer research to facilitate an in-depth understanding of health and social development issues and enhance policy and practice.
Friedman, B. D. (2020). *Community-based Participatory Action Research: It's all about the community*. Cognella Academic Publishing.
  This book explores how to achieve community engagement and community involvement to bring about positive change through research and action. It provides a step-by-step guide to conducting community-based Participatory Action Research (PAR) to understand its transformative nature and encourages readers to think through ethical considerations.
Multiple authors. (2023). Innovative Community Focussed Projects. In Tseliou, E., Demuth, C., Georgaca, E., & Gough, B. (Eds.). *The Routledge international handbook of innovative qualitative psychological research*. Taylor & Francis.
  The series of chapters in this section of the handbook provides an international overview of the latest developments in the field.

Part **3**

Researcher Choices

Being an Objective/ Subjective Researcher

Introduction

As a researcher, you will know that there are different methodological approaches to research and a wide range of methods within them. Objective researchers most commonly use quantitatively-orientated methods and subjective research- ers, qualitatively-orientated methods. The former typically test hypotheses and conduct experiments, the latter explore experiences and their meanings. Mixed methods research combines approaches to reach both depth and breadth in research. Each approach has different requirements and expectations of the researcher and sees their role and their influence on the research differently. In this chapter, we will consider what using each of these approaches, separately and together, means for the researcher. We will consider the overlaps and the differences and how they manifest in practice. We will identify challenges for the researcher using the approaches and discuss how you can overcome these to maintain your role and recognize its influence in your research.

Being an objective researcher

Objective researchers assume that there are independent realities and univer- sal truths that exist outside investigation or observation. Their research aims to uncover these so that they can generalize and replicate the results. These researchers seek to minimize their influence on the research study. In practice, this means that they adopt a neutral stance by, for example, remaining separ- ate from the participants when they are carrying out tests, using other tests to triangulate the results and employing inter-rater reliability strategies so that results across different studies can be reliably compared.

Although objective researchers do not consider themselves part of the research, there are a number of personal aspects and decisions that they should consider when they are conducting research (Partington, 2009; Frost, 2016). We discuss some of these below.

Personal interest and perspective

We have already highlighted the value of having a personal interest in the research you conduct. It motivates you and brings passion and excitement to carrying out your research whatever research approach you are using. Finding a topic to research can come from personal experience, professional interest, community membership and myriad other encounters that leave you with a 'burning question' you want to address. This personal interest and perspective influence the hypotheses and research questions of your study, and the knowledge you already have will shape how you design and carry it out. This means that when deciding on the focus of your research, you will aways bring something of yourself to it. Being aware of this will help to identify, from the outset, the inherent assumptions and particular perspectives that you may be bringing to it (Partington, 2009).

Personal relationships with the data

Personal relationships with the data are formed from the thoughts and feelings you have about it and what you decide to do with them. Data can make you feel angry, frustrated or delighted, as you work with it to interpret and make sense of it. Trying to avoid confirmation bias or imposing assumptions on data requires you to work on your relationship with it, and thus, brings a personal aspect to the research. Decisions about what data you collect, from whom and for what purpose are shaped in part by your expectations of it and what it can tell you. It may be that you have a hypothesis about cognitive processing, for example, and this means you expect the data to confirm or disconfirm this. If it doesn't, you may feel frustrated with the data that you had taken such clear decisions about. This can feed into how you feel about the study, as well as to how you advance it. Objectivity can be threatened by a quest to seek confirmation of answers you think you know already. Rather than hypothesizing and using the research to disconfirm, you may be looking for confirmation of what you think you know. This can mean that you extend the research until you find the answers you want or halt the research when your results seem to fit this. This limits the possibilities of finding alternative explanations and threatens the generalizability of the study results. Being aware of your relationship with the data, and how it changes as the research progresses, brings an enhanced awareness of your engagement with the data and how that may be informing decisions to extend or halt the research.

Personal characteristics and biography

Your personal characteristics influence the study design, research processes and conduct. If you are someone who enjoys engaging with people, you may choose a design that enables you to be more present in the research, such as administering a questionnaire or survey, for example. If you prefer to work with less personal interaction, you may choose to administer online surveys or computer-based tests instead. It is useful to consider your characteristics as a researcher to understand which aspects of research you enjoy the most – perhaps the literature review carried out in the solitude of your desk rather than approaching people to ask them to take part– and how these can shape how you design the study. Knowing yourself in this way helps to both identify limitations you have put on the study and find creative ways to develop it.

Your personal biography can also influence your objectivity as a researcher. Perhaps, you are someone who has always worked with detail or someone who has always thought more abstractly. Your history of working alone or in a team will shape your preferences as a researcher. It is useful to understand the influence of biographical aspects of yourself as a researcher so that you can see how they might have introduced bias or preference into research designs and conduct. Ask yourselves, 'What are the consequences of these for my decisions about research design, conduct and outcomes? How has my prior experience influenced the ways I am seeking answers through research?'

Being aware of how who you are can shape and inform your researcher role is important to enhance the quality and rigour of the research. Although this awareness may never make it into reporting on the research, it is an important consideration that you will take time to reflect on as a successful and responsible researcher.

Being a subjective researcher

Subjective researchers assume that their presence and involvement in the research are central to it and have an effect. They start from the premise that human behaviour and experience are context-dependent and, therefore, recognizing their presence in researching it is important. Participants are seen as having multiple realities, and research seeks to bring understanding and explanation of these by including the researcher subjectivity in conducting, interpreting and reporting the research. Rather than being witnesses to the research, subjective researchers commonly see themselves as co-constructors with the participants of it because of the personal characteristics, experience and decisions they bring to it. The subjective researcher's role is not to remove themselves from the research process but to make as explicit as possible their role in and influence on it. They recognize that they are part of the context in which data is elicited and analyzed and that the context in which they have gathered their knowledge and made assumptions about the world may differ to that of the participants. Being aware of how assumptions and biases, and pre-existing knowledge can influence the research process, from what literature is consulted to gathering data to making interpretations of it, means that the subjectivity of researchers is recognized and that the findings of the research are applicable to the context in which it was conducted. Rather than seeking to generalize, subjective researchers explore the complexity of human experience and seek credibility in their research by evaluating it for trustworthiness. They make transparent as much as is possible by considering their engagement with it (reflexivity) and how this has shaped its direction, data collection and analysis.

Personal interest and perspective

Subjective researchers place more emphasis on being clear and transparent about who they are than objective researchers do and consider how this might inform the research. Being a subjective researcher, therefore, comes with a responsibility for careful reflection on your interest, motivation and perspective on the topic and its investigation. You should think about what has ignited your curiosity in the topic, what you think you know about it and how you know this. Understanding what it is about your own life and personal characteristics that has led you to seek an explanation and more understanding of a topic means you will be better aware of what this means for the research itself. If, for example, you are researching an experience that you have had yourself, you may be challenged to keep this separate from the experiences recounted by participants. Conversely, if you are researching an experience you have not had or conducting research with a community that you are not part of, you must be aware of what assumptions and beliefs you have about them and why you have them. This will help you to put aside assumptions that might obscure intended meanings in participant accounts and minimize the risk of imposing sense-making of your own experiences onto those of others. This means being reflexive throughout the research process.

Reflexivity

Being reflexive means taking time to be self-reflective and to consider this in relation to how you are engaging with the research. This can be hard for researchers not used to doing this because it can feel uncomfortable, sometimes bringing a sense of vulnerability and exposure. It can be helpful to remember that reflexivity does not mean you have to share personal information with future readers. Instead, it means thinking about what such information might mean for the research and how it might impact it. Presenting this, for example, in how you have decided on a research question or in interpretation of data, shows readers that you recognize the subjective influence you bring to the research but that by combining this with systematic data collection and analysis, you are ensuring the research is credible and trustworthy.

Of course, subjective researchers cannot be aware of all the ways they may influence the research – you cannot know exactly how you are perceived by participants and how that influences what they choose to tell you, for example – but you can take time throughout the research process to question and identify ways in which you are a part of it. Keeping a research journal in which you can reflect on research challenges and how you address them can help and provides a platform on which you can ask questions of yourself and the research process. No one else has to see the journal but it will help to maintain an awareness of your role in the study, serve as a useful aide-mémoire when you are writing up the research and provide information with which you can describe your reflexive position in the research write-up.

Personal characteristics and biography

Being a subjective researcher requires a willingness to inquire into yourself and to be honest with what you find. It means being open about your worldview and considering what that can mean for your research. It also means recognizing your reactions to participants and data that may be difficult for you to accept and finding ways to manage such emotions arising during the research. For some, this can feel like an additional and unwanted burden, arousing feelings of vulnerability and concern about undesirable aspects of yourself, but it opens an avenue into a better understanding of your influence on your research practice. In turn, this enhances the quality of the research you do.

If you have had experiences that have prompted your interest in the research topic, it may be painful to re-visit them and their role in motivating your research interest. On the other hand, acknowledging your subjectivity as a researcher can strengthen your determination to benefit others by researching them. Similarly, if you are a person who is simply curious about the world and the people in it, being a subjective researcher can help to focus your curiosity onto a topic that you feel strongly about and want to share with others.

Being an objective *and* a subjective researcher

Researchers who have to be both objective and subjective because they mix quantitatively-orientated and qualitatively-orientated methods have to be both part of the research and stand outside it. Using different methods can mean being open to different worldviews and to what constitutes knowledge – that a universal reality can exist on one level and that understanding or explaining it can be enriched by multiple experiential realities on another. Clearly, there are challenges in being the researcher who works with this combination of approaches. How do you put aside what you fundamentally believe about how you understand the world to investigate it from a different perspective?

One way is to remain focused on the questions you are asking of the data and justifying the choice of each method brought to it. You may want to ask questions such as 'How many?' or 'How often?' alongside questions such as 'How is this understood by or what does it mean for those who experience it?', for example. Of course, these questions require different data, so your relationship with the data sets will also differ. By considering these relationships, you can identify ways in which your worldview influences your acceptance of numbers, ratings and measurements and combine this with how you regard insight into these, which is generated from verbal, visual and other qualitative data.

In doing this, it is useful to think about your method and data biases and preferences. The worldview that mixed methods researchers bring to their research may mean they regard one approach as being of higher status (Bryman, 2007). This influences how the research is carried out and the outcomes reported. It may be that your worldview (and research methods training) mean that you have a greater leaning towards a particular approach or method. It is important to recognize this and make clear the role and status of each method from the outset. Of course, the data analysis in both cases must be systematic and work to address the research question(s) and hypotheses. Your role as the researcher is to make clear the status of each method and try to minimize any personal bias towards a preferred method from infiltrating the research and its write-up.

One approach can be to explicitly prioritize the subjective methods you use in mixed methods research. Qualitatively-driven mixed methods approaches recognize the contribution of the objective method but place the researcher at the centre of the research by using qualitative methods as the driver for it. This enables you as the researcher to maintain your focus on seeking explanations and how you are part of the research. At the same time, by mixing methods at the data-gathering stage, the data analysis stage or at the data interpretation stage, you are aware of the purpose of each method, what knowledge it seeks to generate and how it is relevant to the research question (Hesse-Biber, 2010). Your role at each stage can be considered as you zoom inwards (increasingly subjective) and outwards (increasingly objective) from the research, always with attention to the research question and what role your use of each method has in addressing it.

Being a pluralistic subjective researcher

Pluralistic qualitative researchers combine qualitative methods to carry out research that seeks to understand human experience from different perspectives. In contrast to mixed methods research that combines objective and subjective roles of researchers, researchers doing qualitative pluralistic research always hold the worldview that multiple realities exist. At first thought then, it should not be challenging for pluralistic qualitative researchers to be clear about their role. However, qualitative research, and the accompanying expectation of researchers being clear about their subjectivity in it, have many variations and methods, and these can also differ in epistemological and ontological underpinnings. Whilst most subjective researchers will hold an ontological position that knowledge is not absolute and is generated relative to cultural, historical and societal contexts, questions arising for subjective researchers about how accepting they are of how these contexts are considered. That they vary for individuals is almost beyond contention but whether they are absolute or influenced by factors about people or discourses that pervade them is of concern to researchers with a critical realist view, for example. This, in turn, raises questions about how to access this knowledge. Is it created through interaction (social constructionism) or though cognitive processes (constructivism), for example?

One way to address this as a researcher is to employ different lenses with which to view your research and your role in it. You can employ empirical precision on a micro-level and combine this with descriptive precision

(Hitchcock & Onwuegbuzie, 2022). Being clear about what type of knowledge you are seeking to generate and how you are doing this with each method means that you can foreground a particular epistemological stance whilst also maintaining a different one in the background. Colohan, Tunariu and O'Dell (2012) call this relaxed awareness, meaning that you are always aware of and open to different understandings of how the knowledge is being generated as you conduct the research process. Being clear about which approach to knowledge generation you are foregrounding but that other understandings may also be present in the analysis allows you to ensure rigorous use of each method. Maintaining a journal to record observations, ideas and thoughts about how you are understanding what you are doing and finding allows you to combine all the insights when you write up the study.

A pragmatic researcher combines approaches to what constitutes knowledge and how it is generated by asking 'what works best?' rather than 'what approach(es) to use, when choosing methods' (e.g. Onwuegbuzie & Leech, 2005; Biddle & Schafft, 2015). This does not mean an anything-goes approach to choosing methods but instead enables you to be flexible and collaborative in selecting from the wide repertoire of methods available. Making decisions about which methods to use means researchers must be clear about what each will bring to the research (and what it will not), what research question/hypothesis it will address and how it will be considered in relation to the other methods used. As in pluralistic research approaches, apparent contradictions and tensions across different philosophical underpinnings are seen as enabling dialogue across them, rather than as mutually exclusive (e.g. Clarke, Caddick & Frost, 2016). Apparent outliers are seen as opportunities to consider unanticipated findings, and gaps in the findings of the study can be seen as avenues for further research. Being able to vary how you see and understand the research, whilst also employing rigorous and systematic analysis of the data, means you can be innovative in the research design and acknowledge your subjective understanding and conduct of it. The assumptions brought by each method you employ also drive how you position yourself, and are positioned, within the research. This often means that you are a researcher seeking a more holistic research endeavour based on openness and credibility.

There are also some practical strategies that the mixed methods and qualitative pluralistic researcher can use to facilitate switching between methods, worldviews and philosophical underpinnings. Choosing to use the method requiring a focus on the micro-detail first can generate questions about the macro-detail you may be moving on to. Conversely, focusing first on macro-detail can evoke curiosity about particular aspects of it that you will be excited to explore in more detail with an ensuing focus on micro-details with the next method. Taking a break from the research after completing the use of one method can provide thinking time and an energy refresh that will allow you to prepare for a change of stance in the next stage with a different method. Finally, if you find it really too hard to switch worldviews or methods, then it may be most appropriate to be prepared to compromise on how you originally designed the research. You can highlight that other methods can extend the research in the future, seek training in a new method or look to recruit other researchers with skills in methods that require a different worldview to work with you in a team-based approach to conducting the research.

Being a pragmatic or pluralistic qualitative researcher can also mean collecting many different forms of data. Managing this, whether in one study or across studies can be challenging. The following reflection describes one researcher's strategies for managing quantities of data across different research studies.

Researcher reflection (from a qualitative pluralistic researcher)

I have inadvertently become a hoarder of a lot of data, as my projects typically generate both verbal and visual data in attempts to capture complex life experiences and their psychological meanings. I try to continually move pieces from each project forward, to keep some level of progress across all of these, but at times I find it better to fully immerse myself in one at a time, such as during data collection and analysis, and leaving data from other projects aside during these times.

Researcher positionality

Closely related to the discussions above is researcher positionality. As we have discussed in previous chapters, researcher positionality describes an individual's worldview and the positions they adopt in the research and its social and political context. An individual's worldview includes their beliefs about what is knowable about the world (ontology) and how knowledge can be accessed (epistemology). Positionality has some fixed aspects, such as gender, age, class, race and so on, and other, more fluid, subjective and contextual aspects, such as life-history, personal experiences and political views (Holmes, 2020). It describes the values and beliefs that you bring to your research and includes personal characteristics, experience and demographics that make up your identity. It is key to how you acquire and interpret knowledge about the world. Your position(s) as a researcher will be informed by personal and professional history, motivation for doing the research and what the research process evokes in you. Understanding your researcher positionality is important to the quality of the research because it can give insight into how and why you have responded to participants and what that means for the data that is elicited from them, and why you have interpreted data in the ways that you have. It can challenge your knowledge assumptions and raise your awareness of gaps in your knowledge. Recognizing different positions you take up in the research means that the quality of the research will be enhanced because it can contribute to the explanations of some findings and results as well as biases or differences in the data collection.

How you position yourself, or are positioned by others, informs the ways in which you acquire, generate and interpret knowledge and, arguably therefore, it is important to consider positionality regardless of whether you are using quantitatively- or qualitatively-orientated approaches.

Positionality and the objective researcher

As an objective researcher who regards themselves as neutral in the research process, you will aim to position yourself outside the research. However, this position may be challenged if you find the data personally distressing or are annoyed by unexpected results or with participants who do not fit the assumptions you made when recruiting. Recognizing your positionality and how you have created a position or had one imposed upon you can help identify experimenter bias (when your cognitive bias unconsciously influences the research process) and participant demand characteristics (when participants respond in ways they think you want them to). Objective researchers can regard these as a threat to the validity of the research but an awareness of your positionality, what underlies it and how this can change, can mean that you better recognize mistaken inferences or imposed meanings in the study. By identifying your worldview and how this has influenced your expectations of participants and their responses, you can better understand apparent anomalies in the data and its interpretation.

Of course, it is not always possible to recognize positions you have taken up but acknowledging your positionality, and that you may take up different positions in the research, helps in planning and carrying out the research. For the objective researcher seeking to remain outside the research, this can mean adhering as far as possible to agreed procedures, administration of tests and interpretation of data to help maintain their neutral position. It can also help to consider why you are carrying out the research in the way that you are and what you hope to find from it. It can help you understand why you have selected the literature that informs your hypothesis, who you have sought to recruit to the study and why you think these are the most appropriate people. You can also think about personal experience that has informed your interest in the study and how that might be present in the research process. Attending to these subjective issues not only enhances the quality of your researcher role but also enhances awareness of ethical issues because you will be better able to recognize unintended cues or responses to participants, that, in turn, inform how they position you.

Furthermore, when your objectivity is challenged, you may be better able to recognize ways in which you have responded and what that means for the research. Think, for example, of administering a questionnaire with specified responses and finding that participants elaborate on their answers (e.g. Ryan & Golden, 2006). Understanding why they might have done this in relation to how they perceive you can help with decisions about the relevance of this additional data in the data analysis. Remaining detached from anything you feel passionate about is hard, so accepting that your positionality may fluctuate during the research better prepares you for recognizing and acknowledging it to ensure that the credibility of the research is upheld because you will be more likely to work harder to deliver the standardized procedures and be able to recognize and work through bias, emotion or unexpected responses as they arise. You will seek evidence in the data for your interpretation and present the study outcomes with greater confidence in how you have reached them.

Positionality and the subjective researcher

Subjective researchers recognize and acknowledge their subjectivity in their research. They locate their views, values and beliefs about the research design, conduct and outcomes in it, and by understanding their positionality, they consider and make explicit their engagement with the research process (reflexivity). The importance of this is demonstrated in the increasing number of journals that encourage researchers to include positionality statements with their manuscript submissions (e.g. the *Journal of Social and Personal Relationships*, the *Journal of Engineering Education*). Positionality statements aim to make transparent how author identities relate to the research topic and the identities of the participants, and how these identities are represented in the write-up of the research (Roberts, Bareket-Shavit, Dollins, Goldie & Mortenson, 2020). They seek to demonstrate a commitment to equity and inclusion by raising awareness of privilege conferred by the fixed and fluid aspects that comprise positionality so that researchers are better able to work to flatten power hierarchies in the research process and the research audience can understand the researchers' perspectives.

Recognition of your positionality is also helpful in anticipation of being emotionally affected by the research. It may help you to be better able to see when you are taking up positions of allegiance or collusion with that of the participants and how this affects the research and you as a researcher (Ryan-Flood & Gill, 2010).

There are close links between emotions and how they influence your positionality and, therefore, role as a researcher. Recognizing them and addressing them, whether privately in your research journal or in discussion with peers and supervisors, will mean that you can see more clearly what they have brought to the research. Doing this regularly as the research progresses would mean that its quality is under constant review because its outcome is more likely to reflect your subjective awareness and impact and how it combines with the data elicitation and the systematic analysis of it.

Positionality and the objective and subjective researcher

When researching as both an objective and subjective researcher, as in mixed methods research approaches, it can be difficult to know when to consider your positionality. Striving to adopt a neutral stance can mean that you don't consider what position you are inhabiting, yet you will also be using qualitative methods which require you to be reflexive about your role and impact on the research.

This can be challenging as you switch between the two. However, if you accept that who you are and how you perceive participants and their data, whatever approach you are using, are important throughout the research, then you are free to regularly review ways in which you may have influenced its data elicitation, analysis and interpretation. In your role as an objective researcher, your ability to remain detached may be challenged because of something that has happened or been experienced by you. When your role is of a subjective researcher, it may be hard to put aside the impact of such an experience on your perception of data collected as an objective researcher. Recognizing and exploring your positionality and the ways in which it shapes your researcher role and actions can help you to see how it might have impacted the ways you carried out the surveys, interviews and data analysis.

It is more than likely that you will have different identities provoked or imposed with changes in context, methods and participants and reviewing it regularly can only enhance the quality of the research.

Reflective question

What positionality do you approach your research with? Does this make you a subjective or objective researcher, and what can make that change during the research process?

In the final sections of this chapter, we consider what it means to be working alongside researchers who bring different worldviews to research.

Working in teams of objective and subjective researchers

Being a researcher as part of a research team can be a great way to develop your research skills. There may be more experienced members of the team from whom you can learn new ways of researching, you may be an aspiring team leader who will be able to develop leadership skills, and you may have opportunities to mentor team members who are newer to research than you are. Furthermore, working collaboratively has been shown to enable creativity by combining ideas and by drawing on extant knowledge to develop the research in more novel ways (Uzzi, Mukherjee, Stringer & Jones, 2013). Bringing researchers from different disciplines and research practices also helps with more efficient division of labour (Lee, Walsh & Wang 2015).

However, as with any teamwork, challenges and tensions can also arise, and some of these can be around the different perspectives brought to the research by different team members. On a practical level, the research may be driven by members who share a research approach and who have their own preferences and biases for conducting and reporting research. This can lead to your research approach being less prominent or being given a lower status or your co-authorship of the resulting papers being less representative of your input to the study. In turn, this can lead to frustration and resentment in the team.

One way to think about addressing this is by using the A-B-C model of teamwork (Salas, Cooke & Rosen, 2008). This provides a framework to think about what it means to be a researcher in a team and how to capitalize on both your work and your enjoyment of it. The A-B-C model identifies Attitudes, (shared) Behaviours and Cognitions as essential components to be considered in being a successful team. Attitudes are what team members believe or feel and include openness, trust, cohesion and team viability. Shared behaviours describe what team members do and include communication, collaboration, conflict and leadership styles. Cognitions, which include transactive memory, shared mental models, and information and knowledge exchange, are what team members think and know (Delice, Rousseau & Feitosa, 2019). The three components contribute to the team performance and influence its dynamics, but it is also important to consider how a team develops over time and as it carries out different tasks. In other words, teams have a past, present and a future (Mohammed, Hamilton & Lim, 2008), so it is important to hold in mind and observe how the team develops over time. There may be pinch points when, for example, different methods or approaches are prioritized, and if they are different to yours, then you may have to take a backseat. At other times, your expertise may be called upon to discuss and decide the way forward for the research. Each of these situations will influence how the team works and how your role as an objective or subjective researcher is valued.

It can also be helpful to be reflexive as a team. Rather than considering only your own engagement with the research, it is helpful to consider your relationships with the other researchers and how these shape and inform the research process you are all engaged with. Whilst the relationships can be interpersonal, a reflexive consideration of the team research can open up dialogues about different theories, methods and approaches to the study (Frost, 2016). With a successful reflexive stance, you and other team members can ask each other to clarify their perspective on the research, and to explain how their method is used and how it addresses the research question/hypothesis. Being asked questions about your perspective and method can help not only to make it clear but also to raise any challenges that may be anticipated. Team discussions about the different research perspectives are also helpful in making clear the status of each research approach and method. You will be clear about whether the approach you are bringing to the study is primary, secondary or equal in status to the others and with an open and reflexive team approach, this can also help reduce feelings of incompetency or embarrassment. With effective communication, challenges that arise for individual researchers can be brought to the team before they threaten the research.

Working as part of a team can also highlight differences in perspectives where you may have thought there were none. You may, for example, realize that although you are using the same method as other team members, you are employing it differently, perhaps with a different epistemological approach (see e.g. King, Finlay, Ashworth, Smith, Langdridge & Butt, 2008). Similarly, if you are using different methods within the same methodological approach (such as qualitative pluralistic research), you may think that because you are all using qualitative methods, there will be few differences in how the methods are employed. However, discussion with other team members may show how different subjectivities have influenced interpretations or raised new questions about the research (see e.g. Dempsey et al., 2019). Rather than seeing this as a cause for concern, it can become a strength of the research and influence how the status of the different perspectives brought by subjective and objective researchers is reported.

Objective researchers working with subjective researchers may think that they do not need to involve themselves in reflexive practice as team members. Objective research sees researchers as separate from participants and data, so surely they do not need to consider how others regard it? However, by seeking investigator triangulation (Flick, 2017), the different conceptual perspectives brought by other researchers can develop a theory triangulation, and, in turn, this drives a methodological triangulation. Ultimately, a comprehensive triangulation of researcher, method and data levels can be achieved, enhancing the quality and credibility of the research. To be useful, this approach requires accountability by all researchers in the team so that the outcome of the study is an integrated reflection of the various theoretical backgrounds that have been considered in the methodological planning of the study.

Working as an objective researcher in a team of researchers using both quantitative and qualitative methods enables greater transparency of the impact of each researcher on the research process, links methods to the research question and clarifies the theoretical foundations of the study. Researchers are expected to make clear their individual choices and how they link to the research question, and interpretations reached singly can be discussed and explained within the group. With such an approach, researchers can often be pleased to see that neither objectivity nor subjectivity need to be chosen and, instead, a blended approach can be developed in which aspects of each are considered.

Working as part of a research team can be enjoyable and stimulating. The isolation that sometimes accompanies research as a solo researcher is minimized and a support group is always available (if the team is working well). Opportunities to ask burning questions about the research process are there, as well as opportunities to learn about other methods and to enhance awareness and understanding of your own approach to other interested researchers. You may be working with people from different disciplines or settings and with different stances towards objectivity and subjectivity, so your knowledge of research in other fields will increase. Tensions can be avoided with clear communication and agreements about the role of each team member, and future research can be planned in a collaborative way. If issues such as order of authorship are decided upon early in the process, concerns about this can be avoided (see also APA Guidelines on listing authors in publications: https://www.apa.org/research/responsible/publication) and the focus on the research can continue with the passion with which you entered into it.

Researching alone

Despite the increase in research team approaches (Jones, 2021), there are many occasions when researchers are working alone. This may be through choice or expectation (many academic research assessments require you to carry out research as a solo researcher, for example). Although beginning to change, the nature of learning how to be a researcher often means that you are schooled as either an objective or a subjective researcher. When your research calls for different approaches and methods, therefore, there can be additional challenges in learning and being open to a worldview that you are less familiar with. If time and money are limited, you may have to accept that you will not be able to train in a particular method and, instead, highlight that its use in future research may help to find additional meanings and results in the data. However, if you are choosing, and competent in using, both subjective and objective perspectives as a solo researcher, it can be helpful to think of yourself as a research instrument. This is not to say that you make yourself devoid of feelings and personal characteristics when carrying out the research, but rather that you actively consider them and reflect on what they are bringing to the research. For example, employing a standardized interview means asking questions that were previously decided upon, but the wording and tone in which you ask them may differ across participants, therefore influencing how they respond. Choosing a statistical test to analyze the data may be dependent on those that you know about or what you think you might find in the data (e.g. Carlson & Wu, 2012).

If your research calls for you to be both objective and subjective, you will need to focus your awareness on both how successful you are in remaining detached when administering tests and how you are part of the research when administering interviews. Switching between the two can be easier if you maintain a consistent focus on who you are and how that shapes what you do. You can ask yourself questions such as 'What is my motivation for doing this research? What do I think I already know about the topic? How do I know this?' at every stage of the research to enhance this. It helps you then to understand why you have chosen the methods that you have and why you think each one contributes to the study. It also helps you to recognize the 'human' responses that you have to it, how you have managed them and what they might mean for the research process. As a research instrument, you have the power to determine the shape and process of the research, so know that even when administering different forms of data collection, your presence is constant and influential (see Frost, 2016, for more on this).

Of course, carrying out research alone does not mean you cannot access support and guidance from others. When you are 'in the moment' of doing the research, your awareness of your worldview and positionality is important to its quality, but when you are thinking and reflecting on it, questions and decisions can be discussed and brainstormed with supervisors and peers. This can help to change or strengthen your rationale for doing what you are doing as an objective or subjective researcher and to recognize when your biography and subjectivity may have entered, unacknowledged, into it.

Chapter summary

In this chapter, we have considered various aspects of what it means to be a researcher who sees themself as objective or subjective in the research, working across different approaches. We have identified the value of reflection and reflexivity in all approaches and how these can be practised as an objective as well as a subjective researcher. We have discussed that whilst your reflections as an objective researcher may not make it into the final written report of the research, being reflective throughout the process will enhance its quality and your understanding of what you are doing, and why. When researching as a subjective researcher, you are expected to be reflexive and to make this explicit. When working with team members who have different perspectives on the research, openness and communication can enhance your understanding of their approach and will add to the credibility of the research, as well as its conduct and outcomes.

Further reading

Giere, R. N. (2010). *Scientific perspectivism*. University of Chicago Press.
      This book takes the premise that the nature of scientific knowledge is not absolute and is influenced by the practice and perspective of human agents. It draws on arguments from historians, sociologists and philosophers to show how the acts of observing and theorizing are perspectival, therefore making scientific knowledge contingent.

Scarpinato, K., & Viviani, J. (2021). *Research teams as a goal: Anecdotal tips and action plans for research AVPs and other research administrators*. Independently published.
      This book of practical advice for research team administrators highlights goal creation, team dynamics and communication. Being reminded of the value of these will be of good use to researchers working in teams.

Shan, Y. (Ed.). (2023). *Philosophical foundations of mixed methods research: Dialogues between researchers and philosophers*. Routledge.
      This book includes chapters written by leading mixed methods researchers on the seven main approaches to mixed methods (the pragmatist approach, the transformative approach, the Indigenous approach, the dialectical approach, the dialectical pluralist approach, the performative approach and the realist approach). Each is accompanied by a chapter of critical reflections from the philosophers' points of view. It offers a systematic and critical examination of these positions: a platform to promote a dialogue between mixed methods researchers and philosophers of science.

8 Researchers and Data

Introduction

Data is everywhere. As we go about our lives, we continually collect facts about the world as we see it and process them into information to help us make sense of it. As researchers, we elicit and gather data from participants before subjecting it to tests and analysis to organize it into information that generates knowledge about the topic that we are researching. The raw units that make up data can be numbers, words, pictures, behaviours or anything else that gives us material to work with. The key to the credibility of the knowledge generated from data is the justification for how we elicit and gather the data and what we do with it. These decisions are driven by the reasons for how the data is collected and what we see it giving access to. Broadly, for objective researchers, the facts about the topic under investigation are assumed to be accessed through observable behaviours which, when processed, uncover a universal reality. For subjective researchers, insight into the topic is gathered by taking context into account and is accessed by asking people about it. Accounts are analyzed to generate knowledge that can explain what the data means.

For all researchers, therefore, eliciting data needs to be carefully considered so that the rationale for the type of data collected is in accordance with the underlying assumptions about how it will be used to address the hypothesis or research question. In turn this raises questions about what it means for researchers to gather different forms of data. In this chapter we will discuss some of the most common forms of data and how it is collected, and consider it as an experience for the researcher.

Statistical data

Statistical data is data that has been collected or generated by statistics in the process of statistical observations or data processing (European Commission Collaboration for Research and Methodology for Official Statistics, https://cros-legacy.ec.europa.eu/content/statistical-data_en). As a researcher working with statistics, you are likely to be either gathering your own measurements and observations for analysis or consulting existing databases about populations or subpopulations. You will choose how to analyze the data in accordance with what it is you want to know from its processing. It may be that working with such datasets alone will be sufficient for your inquiry, but it is not uncommon to look into integrating it with new data that you have gathered yourself and with other extant information. Think, for example, of how you might gather data about shopping trends on a particular high street. You may have observational data about how many people visit which shops and what they buy. You may also want to know which days are the busiest for the shops, what the effect of the weather is on shopping activities and why people choose that street to shop.

Gathering such information introduces contextual information and potentially introduces new variables and outliers, about which you will have to make decisions (e.g. whether to adjust for them, see them as confounding, or ignore or include them). These issues challenge the apparent objectivity of statistical data because it is you, the researcher, who makes the decisions, and these will be based on your prior knowledge and assumptions. Considering yourself as an objective researcher will mean you seek to eliminate subjective inferences from your research but this can risk self-censorship and lead to potentially important omissions from the research (Gelman & Hennig, 2017).

One way to address this is to reconsider the terms 'objectivity' and subjectivity' so that they are not seen as being in opposition but as each including elements that can be combined to complement the goals of the research. It can be useful, therefore, to think of 'objectivity' as having transparency, consensus, impartiality and correspondence to observable reality. 'Subjectivity' can be understood as having awareness that multiple perspectives exist and that data is context-dependent. Not all elements of each will always be included when working with statistical data but an openness to including those which are relevant to the data can enhance the reliability and quality of the research because they provide clearer audit trails and understanding of bias in decision-making about statistical tests.

It can also help to think of yourself as a research instrument rather than simply a witness to it (see also Chapter 7). Considering how your objectivity is open to compromise and recognizing where you may have advertently or inadvertently influenced the choice and administration of tests mean that you can better recognize breaching of the objective-subjective divide (Frost, 2016). By acknowledging that it is not only the test that is the research instrument but also you, the person administering it, as well as generating and interpreting the results, you can elicit more meaningful and useful data (Frost, 2016). This can be particularly pertinent for community researchers but is applicable in most research settings.

Questionnaire and survey data

Questionnaires and surveys can consist of only categorical responses or a combination of categorical and free-text data. Your role as a researcher working with questionnaire and survey data means that you need to make decisions about the status of the forms of data collected, e.g. will categorical data support or be prioritized over free-text responses? You should also consider how the data has been elicited (online or in-person), where it has been collected (public place, workplace, participants' homes, etc.) and the participants from whom you are collecting it (random sample or representative). Each decision is important to the data itself and to how the knowledge generated from it is understood. For example, online respondents may take different amounts of time to consider their responses than those responding in-person. This can mean that the responses themselves may have a participant bias or have been changed before being submitted. In-person respondents may have perceptions of the people gathering the data: were respondents wary about being honest in their replies for fear of judgement or impact on the service they receive? Were the administrators seen as insider or outsider researchers? Was it made clear to participants who will see the data and how it will be anonymized? These considerations can help make sense of data that does not seem 'to fit'.

If you are using a standardized questionnaire or survey from which you are going to generate a second phase of the study, you need to be clear about what role each data set is going to play in testing the hypothesis or addressing the research question. If you are using a survey that allows for free responses, then you need to have a rationale for this and a design that makes clear what the purpose of each form of data will be.

Careful training in data gathering means not only that researchers are clear about what is expected in the administration of the data collection instrument, but also allows for the human characteristics that may influence how they do it (Pezalla, Pettigrew & Miller-Day, 2012). It is important to consider how you are presenting yourself and the purpose of the research when gathering data. Thinking beforehand about possibilities of respondents giving extraneous information or asking you questions about your interest in the data can prepare you for unexpected opportunities to extend or enrich the data during the process. If you are administering the questionnaires or surveys online, it can be useful to also gather information about respondents that may give you enhanced information about the context of its collection. Finally, you can also think about why you have made the choice to administer online or in-person. For example, is it simply a resources-driven decision or is it also about your desire/reluctance to engage directly with people in different settings?

The choice to use questionnaires and survey data collection methods will, of course, be decided in part by the research focus you have chosen and the knowledge you want to generate, but it is also useful to consider what making this decision means for you as the researcher. Understanding what you know and believe (positionality) and what you do with it (reflexivity) means that you can reflect on how you develop and shape the research conduct. Thinking through questions such as the meaning of the topic to you, how and why you have selected the pool of participants from which to recruit and why you are interested in patterns of commonality rather than individual experience can help you to see where you might have perpetuated bias or pursued a personal agenda in your research (Jamieson, Govaart & Pownall, 2023). Such reflection can be aided by, for example, pre-registration of sample size and characteristics, and recruitment strategies. This makes more transparent what you plan to do and opens it up to questions about exclusion or bias.

Interview data

Interviews are used to gather data by asking people to tell you about experiences they have had. They range in style from standardized interviews, in which set questions are asked in a specific order, to semi-structured interviews, which use a combination of few open questions (to enable detailed responses) and follow-up probe questions (to enrich responses and follow the direction each participant takes) (e.g. McAdams, 1993), to single-question biographical interviews (e.g. Wengraf, 2004), which aim to enable the participant to talk freely and uninterrupted. To some researchers, interviewing people is the best part of the research, whilst for others, the thought of sitting down with strangers and asking them questions is a daunting prospect.

Part of the reason for these feelings about interviewing is the expectation that a better interview will result from developing a rapport with the interviewee. This means helping participants to feel at ease so that they feel more able to talk openly to you. The onus is on you to develop a style of interviewing that will elicit the type of rich data that you want for your research. This can mean avoiding appearing to be judgemental or critical, trying not to close down participant responses, allowing for silences and generally working to maintain the flow of the interview. For researchers who are natural conversationalists, it can be hard to remain in the background of the interview or not to interject with responses that lead a participant in a different direction to the one that they had intended. For researchers who are shy or nervous in interviews, there can be a risk of the interviews becoming stilted. These can mean they end abruptly as both you and the participant rush to get away from the situation.

Developing a rapport can be hindered if you are making notes or looking at the questions you want to ask. Instead, memorize them, focus on listening to the participant, and make sure you record the session (with permission). If your interview is a standardized one with too many questions to memorize, and you need to write down participant responses, make sure you explain that this is what you are doing and offer to let the participant see what you have written.

With the development of technology, there are alternatives to conducting in-person interviews. Video interviews can be close to the experience of being there together, as long as both you and the participant are familiar and comfortable with using the software. There are functions to record what is said. There can be issues with privacy, however. Make sure you and your participant are in an office or room where neither of you will be interrupted, and that extant noise is minimal Distractions caused by people coming into the room, or noise such as construction or traffic can be difficult to overcome. It is also important to do all that you can to maintain confidentiality. This means setting a password for the site and making sure the interview is stored appropriately when it has ended.

Telephone interviewing can be as effective as in-person interviewing (Holt, 2010). It may be easier for participants to talk about sensitive topics by telephone, but the detachment brought about by not being able to see the interviewer/participant can also hinder the rapport. By focusing on tone and intonation, you can use your skills to adapt your style. Like video interviews, telephone interviewing enables access to participants in geographically diverse places or in hard-to-reach places such as conflict zones or those with unstable internet connection (Saarijärvi & Bratt, 2021). It is important to check that the connection is a good one and also that you and the participants are comfortable and confident using the telephone. There is a risk that the person you are talking to is not who they say they are, but establishing trust in the recruitment stage can help to minimize this.

Online and email interviews raise additional challenges in developing an interaction. It is important to carefully consider the language used in asking questions and to be aware that written words can be read in different ways. Remain aware that language used by the participants can also be misread by you. One way to mitigate this to some extent is to use emojis to symbolize intended meanings of potentially ambiguous wording. Setting up a chart of emoji meanings that can be shared with participants beforehand further aids clarity. When analyzing interviews conducted online, it is important to remember that they can differ to in-person and other styles of interviews conducted in real time because participants may take more time to reflect before responding. It is useful to make a note to yourself when some responses are slower in coming than others – it may point to a concern in answering your question.

With reflection, you will be able to better understand the experience of developing rapport with participants. Were there some you warmed to more than others? Why was this? How did this manifest in your interview interactions? Were online or telephone interviews conducted by your choice or for your convenience, or for the participants' convenience? How comfortable are you with technological interactions rather than in-person ones in your non-researcher life? What did this mean for how you presented yourself and communicated with the participants? Were you aware of their level of dis/comfort and what it might mean for how they responded? Questions such as these will enhance your reflexive awareness so that when it comes to analyzing and interpreting the data, you will be better able to see where and why you may have influenced the data elicitation.

Internet, mobile and technological data

The internet provides a rich and extensive source of data for research and it can be collected as a form of 'online ethnography'. In addition to providing quantitative data that can be used in statistical and trend analysis (see sections on 'big data' and 'statistical data' in this chapter), the many fora, blogs, virtual worlds and mobile communication sites, such as Facebook and Instagram, allow for rich insight into online human communication and meaning-making. It can enable the study of interaction between communities and cultures not otherwise accessible to you (Kozinets, 2002). It can be less costly, less time-consuming and less intrusive than seeking out participants and groups from whom to elicit data in person. Online ethnography can range from simply downloading images and textual data for analysis, through to engaging participants online by joining communities and seeking interactions for collection as data.

Netnography (Kozinets, 1999, 2002, 2015, 2019) is a term that describes a particular approach to gathering data from the internet. It emphasizes the importance of humanism, paying attention to details and contexts of human stories, and human understandings of people using technologies. It requires researchers to engage as fully as possible with online participants (Kozinets, 2019). Netnography can focus on gathering data from particular fields, such as virtual worlds, or on particular approaches, such as discussion fora, and can gather both textual and visual data.

A netnographic researcher, therefore, plans to maximize their engagement with participants (perhaps, by following up data collection from sites with an online focus group) and includes time for reflection and introspection in the research design. As an ethnographic researcher using online data, you will regard yourself as a research instrument recognizing, as far as possible, your influence and role in interactions, observations and reflections. This can mean making extensive fieldnotes and written recordings of interactions between yourself and participants to aid data analysis and interpretations of how understandings of the world are created online. The online and offline worlds can be blended, and the process and outcomes are close to traditional ethnographic approaches where the researcher is part of the community they are researching.

It is useful to consider what this approach can mean for you as a researcher. It requires you to engage with both online data and human participants, meaning you have to draw on technological skills as well as person-to-person communication skills. Which of these are you most comfortable with? How can you ensure you have the appropriate expertise in each? It may also raise questions about how you approach the generation of the different data sets and how you regard the two forms of data gathered. How will your worldview inform the status of each and their integration? As a researcher, you will always have to consider the epistemology and ontology underlying your research, and it may be that with this approach, you will bring different beliefs about what written data and spoken data can tell you. That, in turn, may mean your relationship with each form of data will be different.

As an online researcher, you will need also to make some ethical decisions. Are you going to collect the data *covertly*, by watching fora and other sites, *overtly*, by joining and posting on the sites to generate data for collection, or by using a mixture of the two? In the latter case, you may initially join a community to understand its purpose and structure and get to know other participants, and at a later stage, declare yourself as a researcher to gather more data (e.g. Álvarez, Sintas & Martínez, 2012). Asking yourself how the risk of being exposed and confronted about this can help with your decision to use this approach (Podschuweit, 2021).

You will also have to make some decisions about the ethical position you are going to take when gathering online data. Lee (1993) has defined three ethical stances:

- Absolutists stance, where covert observations and data collection of private data are fatally compromised
- Sceptical stance, where a positive justification can be made for collecting the data this way
- Pragmatic stance, which recognizes the potential ethical difficulties of covertly collecting data but respects the need to protect the rights of participants and the researcher obligation not to cause harm to them as a result of taking part in the research.

Considering which stance you will adopt can make the decision about using online data easier.

Gathering data as a covert researcher can be pertinent if you are researching sensitive or rarely talked about topics, such as cosmetic surgery (e.g. Langer & Beckman, 2005) or illegal drug purchase (Fittler, B sze & Botz, 2013). Whilst netnography may be an efficient and useful approach for this, it is important to consider that the topics themselves cannot be assumed to be fully objective and may evoke feelings and emotions in you. Collecting data from the internet may seem objective, but the topic may provoke distress or anger in you, and the lack of human interaction may leave you feeling powerless or frustrated at not being able to show concern. Having a reflexive relationship with your research can help to anticipate and manage feelings and emotions arising from it.

Reflective question

Are you a researcher who prefers gathering data in-person or online? What is it about you that leads to this preference? How does it inform your research?

Secondary data

In this section, we are going to consider what being a researcher who uses secondary data means. The next section on Big Data looks more closely at being a researcher who uses data collected nationally or by organizations and institutions.

Secondary data can be data you collected yourself in an earlier study (the parent study), data shared formally or informally by other researchers, or data held by institutions and public bodies that grant you access to it. Depending on what you want to know from the reuse of data, the raw data can be re-analyzed or the results of original studies can be used for meta-analysis. It is an increasing requirement of funding bodies that data is made available for reuse by other researchers, and many countries have national databases of data for researchers to access (e.g. UKdataservice.ac.uk, the Danish Data Service (https://en.rigsarkivet.dk/), the Irish Social Science Data Archive (https://www.ucd.ie/issda/); also see World Bank Databank for a collection of databases from around the world (https://databank.worldbank.org/)).

Apart from the obvious advantages that using secondary data offers, such as saving time and labour in gaining ethical approval, recruiting participants and gathering data, the reuse of research data reduces the likelihood of subjecting potential participants to 'over-researching' (Willig, 2021). As a researcher thinking about using secondary data, however, an important question to ask yourself is 'Why am I choosing secondary data rather than collecting new data myself – is it to save me time or to protect participants from being over researched?'

Secondary data can be used to ask new research questions (although there can be ethical considerations in doing this – see later in this section) or to find additional meanings and results to those of the original study. It can be used to dis/confirm evidence from original studies so that results can be extended and generalized more widely. It offers a way of conducting longitudinal research in a shorter timescale. In qualitative research, secondary data analysis can facilitate theory development because combining data from different studies enables the qualitative researcher to move the research beyond the specific context of each study (Morse, Cheek & Clark, 2018).

However, working with data that has been collected by someone else means that you will be unaware of context and study- or researcher-specific nuances of the data collection (Branney, Reid, Frost, Coan, Mathieson & Woolhouse, 2019). Conversely, if you were involved in the parent study, there may be a risk of carrying over bias from it to the new study. It may also be that identifying details such as participant demographics and location have been redacted so you may not know details of the groups or subgroups and whether they are relevant to your research. Working 'blind' can introduce confounding variables or other aspects of the data that you are unaware of and which can lead to invalid outcomes. As a researcher using secondary data, you may not know the level of non-responsiveness or selection error. This can mean that the representativeness of the data is compromised and that the interpretations you make from it include and perpetuate this bias.

Similarly, it is important for you to know whether the data you are planning to work with is a summary of the original data set (usually this is what is published in journal articles) or the full data set, so it is important to go to the research source or data archive to access the full raw data set before making your decision to use secondary data.

You can go some way towards addressing these challenges with some practical steps such as seeking out information about the validity of the data – either from documents included with the data set itself or by looking further into publications and other documents relevant to the original research study. It can be useful to ask questions of the data set about its original purpose, the time period in which it was collected and who collected it before deciding whether to use it. Finding out this information can help to contextualize the data so that you can consider it in relation to the research questions you now want to ask of it. This can add time (that you may have thought you had saved by using secondary data) to your data preparation but it will increase the rigour of your study because you can be more sure that the data is relevant to the hypothesis and/or research questions you want to test/ask.

In addition to secondary data's validity and quality issues, it is important that researchers consider ethics when reusing data. Consent may have been given by the participants of the original studies for their data to be used to conduct research into a particular focus of inquiry by particular individuals or organizations, but consideration may not have been given for its reuse by others with different foci. There are moves now to include consent for participants' data to be used by other researchers to be asked for in the Consent Forms for the original studies. Sometimes, there are caveats to this, such as conditions that anonymity is maintained, that data can be reused in one form and not another or that participants can ask for parts of their data to be excluded from future reuse (Branney et al., 2019). This is important to bear in mind if you are planning further research from the secondary data research so that you ensure you seek appropriate consent from participants.

Whilst these considerations go some way towards addressing ethical concerns, they do not address the issue of researchers imposing unintended meanings on data because of a lack of knowledge about the context in which the data was originally collected. Branney et al. (2019) propose bringing a reflexive consideration to both the original dataset and its reuse, to include a specific focus on the context of each. Qualitative researchers using secondary data should collate as much information as possible about the context of the original study and its participants and also consider the context in which they are carrying out the secondary analysis and interpretation. This means identifying, as far as possible, the impact of differences in time, location and culture amongst other dimensions of context, alongside reflexive researcher engagement with the reused data, so that the secondary researcher's role, impact, expectations and motivations are made clear. These can be methodological considerations (e.g. what does it mean to be analyzing data using a method different to the one originally used?), epistemological considerations (e.g. what kind of knowledge are you looking to generate from the secondary data?) and personal considerations (e.g. what do you bring to the data that is the same as/different from the researchers who originally gathered it?). It may not be possible to fully answer these questions but holding them in your mind as you re-analyze data and reflecting on what they may mean for what you are looking for and what you find in secondary data will enhance the quality of the study and go some way towards doing this in as ethical a way as possible.

Big Data

The concept of 'Big Data' has been in use in research since the term was first coined in the mid-1990s. It referred then to the handling of massive datasets (Diebold, 2021) but the definitions are still evolving. The European Commission defines Big Data as:

> Large amounts of different types of data produced from various types of sources, such as people, machines or sensors. This data includes climate information, satellite imagery, digital pictures and videos, transition records or GPS signals. Big Data may involve personal data: that is, any information relating to an individual, and can be anything from a name, a photo, an email address, bank details, posts on social networking websites, medical information, or a computer IP address.
>
> (European Commission)

However, a study that asked researchers in psychology and sociology in Switzerland and the USA found that there was little agreement amongst them about the definition of Big Data and that most referred to it in terms of practice (such as data processing and data source) (Favaretto, De Clercq, Schneble & Elger, 2020). These researchers concluded that the term itself is vague and instead see 'Big Data as a shifting and evolving cultural and scholarly phenomenon—or a cluster concept that includes a plethora of sophisticated and evolving computing methodologies—rather than a clearly defined and single entity, or methodology' (p. 17).

The '3Vs' comprise three key traits of Big Data: volume (enormous quantities of data), velocity (created in real-time) and variety (structured, semi-structured or unstructured) (Kitchin & McArdle, 2016). However, some researchers argue that the Vs models are too technical and do not fully capture components of Big Data. They have developed the Ps model which includes the personal, political and predictive (Lupton, 2015). Others criticize both the Ps and Vs models as they only describe a broad set of issues associated with Big Data, rather than characterizing the ontological traits of data themselves (Kitchin & McArdle, 2016). They contend that volume and variety are not key characteristics of Big Data and only velocity and exhaustivity are.

Still others define Big Data as data in the range of exabytes and beyond (Kaisler, Armour, Espinosa & Money, 2013), and with the rapid expansion of technology, there have been debates about other elements that make Big Data Big Data. These include value and variability, and exhaustivity (comprising an entire data set rather than a sample from it) (Mayer-Schönberger & Cukier, 2013), relationality (containing common fields that enable the conjoining of different data sets) (Boyd & Crawford, 2012) and scalability (can expand in size rapidly) (Warren & Marz, 2015).

So, as a researcher using Big Data, the first question you must ask yourself is what you mean by Big Data. The second question might be how are you going to access it? And the third, what are you going to do with it?

To answer the first question, the best advice is to read the literature and to talk to your fellow researchers. It is unlikely that you will be researching with Big Data on your own so it will be important that the stakeholders involved in the research (the organization, Research and Development office, the commissioners of the research and so on) have an agreed definition from the outset. This will help to clarify the research design and rationale, as well as introduce an agreed language with which to write up and disseminate the research.

The question of access also requires thought from early on in the process. The quantity of data involved in Big Data research is massive, so there are challenges not only in accessing it but in storing it too. This is made more complicated by the velocity with which it is produced. The issue of accessing Big Data is one that puts off researchers who do not work in institutions that gather it from their own digital platforms. Furthermore, questions of privacy and ethics mean that many platforms are not willing to give access to their data. Some universities pay for access, but the majority do not. There are, however, other ways to gain legitimate access. Data may be scraped from public sites (such as Reddit) using automated procedures, or payment can be made to external vendors for data such as tweets. With such access, participants can be directly recruited to contribute their data, although given the number of participants you may be looking for, perhaps tens of thousands, this can be a costly and challenging approach.

Perhaps a more manageable approach is to use an online survey tool such as that offered by Qualtrics. Widely available, this enables researchers to use built-in questions to construct surveys which participants can scroll through in ways similar to how they would traverse an online website (Adjerid & Kelley, 2018). Qualtrics also includes features that allow researchers to collect data on the respondents' behaviour in these environments, such as how long it takes respondents to answer questions, how many clicks they make and how long they spend on a page. Such features enrich the researcher's ability to collect data in terms of quantity as well as quality, over manageable periods of time (ibid.).

However, there still remains the issue for Big Data researchers as to how to prepare the data so that it is meaningful for analysis. Gathering such large quantities from unknown participants and a range of online platforms introduces concerns about the use of automated bots, and having clear systems and capabilities for its storage can be costly. Given the quantity of data and the speed at which it is collected, it is not feasible to have human research assistants to 'clean' it and, instead, requires highly technological expertise. Whilst this is being addressed through the development of web-scraping strategies (e.g. Landers, Brusso, Cavanaugh & Collmus, 2016) and tool kits and software packages (such as Gentry, RSQLite and Artistic), researchers in this field need to have knowledge of the technology and how best to use it. Researcher knowledge can be enhanced by applying the computing skills they have to employ software packages to carry the load of implementing advanced techniques (Chen & Wojcik, 2016). However, for some researchers, it may be important to consider how proficient their computing skills are and whether, in fact, needing to install and use software packages is a limitation to use of this approach.

Related to accessing the data is the question of how much data to gather or, to put it another way, quality vs quantity of data. As a researcher working with Big Data, you must make decisions about whether increasing the quantity of data you have will lead to the perfect explanation of the phenomenon you are interested in. Or whether, conversely, you choose not to increase your access to more data and instead to draw careful conclusions from a small quantity of high-quality data (Kaisler et al., 2013). For example, you may be analyzing trends rather than database systems, so you can ask yourself what level of precision is needed in understanding the large quantity of data. Considering beforehand how much time you will have to carry out the study can help in deciding whether to carry it out.

It is important to end this section with a consideration of the ethics of working with Big Data. It may be that you have access to it, but Big Data platforms may own it. Those who contribute the data might argue it is theirs and they need to provide consent for it to be used in research. Think about some of the arguments that have been presented in courts over the ownership of and access to Facebook data following a death, for example.

Kaisler et al. (2013) suggest four challenges that should be considered by researchers using Big Data:

- When does the validity of (publicly available) data expire?
- If data validity is expired, should the data be removed from public-facing websites or data sets?
- Where and how do we archive expired data? Should we archive it?
- Who is responsible for the fidelity and accuracy of the data? Or is it a case of user beware?

Whilst, as yet, there are few definitive answers to these questions, researchers will do well to be aware of them and make decisions about what data they access and what they do to maintain their responsibility in research for it. This will enhance the researchers' use of Big Data in valued and justifiable ways.

Visual data

Visual data is data that is non-linguistic. It can include photographs, drawings, paintings, objects and film. The data can be provided by participants or the researcher, or generated together. Visual data can also play a role in textual data generation. Asking participants to bring photos or drawings they have chosen to represent an experience to an interview (photoelicitation) can enable them to have more agency over what they talk about and, perhaps, use the image to express what they cannot or do not want to say with words. Photovoice seeks to promote social justice and change, so the researcher might, for example, use existing visual images to garner meanings that they have for participants. This can be useful, for example, when researching social media images to understand their impact.

As a researcher, and depending on the research question, you will be interested in how the participants describe or use the image, or the meanings they ascribe to it. This can be particularly useful if you are inquiring into a topic that may be sensitive or hard to describe (such as spirituality (Majumdar, 2024)).

For researchers contemplating gathering visual data, therefore, there are important questions not only about the form of the visual data but also their role in its elicitation and interpretation. For example, if you are planning to ask participants to bring their own artefact, what role will you give it? Are you asking them to bring something pertinent to the focus of the research because you plan to use it as an 'icebreaker' at the start of the interview or will you regard it as a way of giving meaning (latent or manifest) to the participant's accounts? If you are bringing the artefacts, how will you decide on what to bring (directly representative of the topic or a metaphorical representation of it?) and which out of a selection should you use in the interview? These questions mean that you must consider the epistemological stance(s) that you will be bringing to the visual data collection and analysis, remembering that visual data can be analyzed in a number of ways. Using content analysis, the frequency of different aspects of the images can be quantified and each can be used to provide descriptive insights into, for example, preference trends or behaviours. Visual data can also be used for further analysis (quantitative or qualitative) or for triangulation. Similarly, visual images can be analyzed qualitatively to contextualize quantitative data. The philosophical underpinnings you ascribe to each method will inform the generation of the knowledge you are seeking.

In making your decision to use visual methods, it can be useful to reflect on whether you are someone who connects more with visual or textual representations. What does your preference tell you about how it enriches or provides meaning for your own experiences? As a researcher, knowing this about yourself can raise your awareness of how you may understand participants' use and explanation of their artefacts and help in your involvement with them in creating and describing them and their meanings.

If you plan to be a researcher who is part of the production of images with participants (photoproduction), you may be drawing, painting, photographing or filming with them as part of the data generation process. This means that you must consider your role in deciding which and how the resultant images are included in the research write-up. Gregory Bateson and Margaret Mead (1942) were early proponents of filmmaking and photographing of participants when they researched a comparison of adolescent behaviour in Samoa and Papua with adolescent behaviour in the USA. Although a vast dataset of 25,000 photographic stills and 22,000 feet of film were gathered (Jacknis, 1988), critics have highlighted that in selecting what to present to the wider audience, Mead introduced a subjectivity that contradicted the apparent objectivity of the data (see e.g. Shankman, 2009) and risked misrepresenting the Balinese communities. The data in the study were turned into two books, each featuring many examples of the visual images. However, there are questions about what directions were given to participants in some of the photos, leading to concerns that some of them may have been staged for the researchers. This highlights the importance of transparency from researchers to provide as much clarity as possible about how the study was carried out. Furthermore, it is argued that the editing of the films may have contributed to misleading information, for example, trance dancing being shown of shorter duration than it was.

These are important considerations for researchers using visual data, and these considerations also alert us, in these days of social media and AI, to the risks of photo and film manipulation by others when it is disseminated. These critiques of the methodology and how it is used to represent other people and their experiences provide us with salutary lessons about the power that researchers can hold in explicit co-construction with participants (see also Brody, 2021, for consideration of the challenges of co-constructing maps with participants).

Researchers are increasingly innovative in their use of visual data. One study (Hansen & Flynn, 2015) photographed graffiti on the same wall every day over a period of 36 months, to understand the messages it created in the ways it was altered by others over time. They found a narrative that told its own story but also a visual dialogue that existed between the image and other artists, observers and 'mark-makers' on the image:

> **Researcher reflection** (from Hansen & Flynn, 2015, p. 30)
>
> *Longitudinal photo-documentation allows us to make visible, for subsequent analysis, the dialogue amongst artists, writers and community members, with each party showing their understanding of the prior work on the wall via their own contribution to the 'conversation'*

Mobile methods

Mobile methods are methods that gather data whilst moving (e.g. walking or rolling) with participants. Data gathered using mobile methods can be qualitative (e.g. conducting interviews) or quantitative (e.g. counting frequency of interactions with other people). They can be gathered through interactions between researcher and the participant or by the researcher equipping the participant to gather data themselves (through audio or visual recording, for example). Mobile methods are used to try and get closer to the participants' everyday practices and lifeworlds, both by experiencing them as a researcher and by seeking to understand more about how participants experience it *in situ*. They move researcher and participant out of the sedentary laboratory or interview room into a more relevant and dynamic context which can help not only how participants talk about it but also enrich insight into the experience. Mobile methods can be used to investigate questions ranging from how people perceive their neighbourhood, nature or buildings, to how policy makers, planners and designers can develop areas, communities and architectural sites.

As a researcher collecting data using mobile methods, you will have an interest in people's locations and journeys. These may be in everyday journeys such as commutes or shopping trips, to migrant journeys, to spread of diseases. Mobility can be seen as a symbolic and material connection that serves the human will to connect with places (Sheller & Urry, 2006). Engaging with participants as they move through them can give valuable insight into their perspective of the world and how embodiment informs their emotions and sensory experiences (Thrift, 2016; Hein, Evans & Jones, 2008). It also helps to situate participants in relation to other people and their own multiple identities and roles (Hein, Evans & Jones, 2008).

So, what does it mean to be the researcher collecting data this way? The rise of walking/rolling interviewing sees researchers accompanying participants on their journeys, interviewing them as they travel together. An obvious question to ask is whether your presence influences how the journey is perceived and how participants respond to the questions. On the one hand, the interview may be more informal than one conducted in a static location, allowing you to observe interactions with other aspects of, and people in, the environment. It can enable you or the participant to open up discussions about issues related to the locale as you encounter them. On the other hand, your presence on an everyday journey that is usually made alone or with other, known, people makes the journey a different one. This affects the experience of it, perhaps with a focus on you rather than on the sounds and sights that the participant may usually observe, or perhaps with a concern on the participant's part that there are certain aspects of the journey that you want to hear about and not others. Through your presence, you may be intruding on the private stream of perceptions, emotions and interpretations that participants usually bring to the journey (Lee & Ingold, 2020). Therefore, the experience being recounted cannot be taken as a direct representation of how the journey is usually experienced.

To address this somewhat, you can equip participants with a camera to film elements of the journey that they consider significant. If these are not discussed during the interview itself, the images can be included in your analysis of the interview later on.

Another way to address this issue is to remove the 'planned' nature of the journey and instead adopt a 'bimbling' approach (Anderson, 2004). This is aimless walking, often at places of protest, in which the researcher makes field notes as they travel with the participant and uses these rather than an interview to interpret the meaning of the journey. The researcher is positioned somewhere between an academic and an activist (Soja & Thirdspace, 1996). The data gathered this way is neither mapping and measurement data, nor theory building, but instead seeks the lived experience by paying attention to elements of both of these.

To go further, you may decide as a researcher to remove yourself entirely from the physical context of the data-gathering and instead equip participants with technology to record their experiences audio and/or visually. Advances in mobile phone technology and go-pro cameras mean that participants can be agentic in recording the journey without the interference of a researcher presence and provide data for you to analyze later.

With each mobile method come practical and resource implications. It is important to consider your own and the participants' ability and capacity for mobility. How long are you/they able to walk/roll and talk for? What if the weather is challenging? How and when should breaks in the walk/roll be offered or taken?

Sound can play an important part in understanding the embodied experience of a journey and it can be useful in developing a more holistic insight into the experience. What if traffic or construction noise intrudes to a difficult degree? How will you deal with the noise of crowds? If the journey is in a car, how will you manage radio noise or a malfunction in the car or even feeling unsafe with the participant's driving ability? If the interview is a roll-along one, how will unexpected access restrictions be navigated?

Finally, it is important when working with data collected from mobile interviews to consider your own experience of it and to reflect on how you positioned yourself. If you are aiming to collect quantitative data, how did you decide what to measure or map? What biases may have been inherent in your choice and use of equipment (used by you or by the participant)? How technologically competent are you and/or the participant to ensure the data is preserved for later analysis?

The recent rise in mobile methods shows the promise they offer in bringing researchers closer to participant worlds; it is important, though, to consider how you are gathering the data and what epistemological underpinnings you are ascribing to it. It may be that you want to complement it with follow-up interviews conducted in the more common interview (static) setting in order to triangulate or enrich what has been noted and recorded on the journey. Or it may be that you choose to represent the outcome of the data analysis in less conventional ways, perhaps making another film by using those collected from participants.

Chapter summary

In this chapter, we have discussed several forms of data and considered the benefits and challenges for researchers collecting and analyzing it. As well as the more commonly used data collection methods, such as surveys, questionnaires and interviews, we have considered availability and accessibility of technological data collection. We have discussed Big Data research and using social media for data-gathering. We have also discussed mobile methods of data collection, which offer insight into meanings and experiences of places in physical space and time. Ethical considerations and challenges of each form of data have been highlighted throughout. In Chapter 9 we discuss the role of equality, diversity and inclusion in research practice.

Further reading

Castellani, B. C., & Rajaram, R. (2022). *Big data mining and complexity*. Sage Publications Ltd.

This book offers a critical introduction to data mining and Big Data. It includes multiple case studies and examples, to illustrate key terms and concepts relevant to the use of social media in quantitative research, a complexity science perspective to review data mining and Big Data potential and limitations, and explores challenges of developing and managing a Big Data database.

Humberstone, B., & Prince, H. (2019). *Research methods in outdoor studies*. Routledge.

This book appraises established and cutting-edge approaches to outdoor studies by examining key methodologies, themes, technologies, ethics and research dissemination.

Saris, W. E.. & Gallhofer, I. N. (2014) *Design, evaluation, and analysis of questionnaires for survey research*. Wiley.

This book analyzes the important decisions researchers make throughout the survey design process. It covers the essential methodologies and statistical tools utilized to create reliable and accurate survey questionnaires and features an expanded explanation of the usage and limitations of SQP 2.0.

9 Equity/Equality, Diversity and Inclusion

Introduction

For research to be as beneficial to as many people as possible, the researcher must consider how inclusive it is. The implications of unintended or unrecognized exclusion can have powerful detrimental effects on those who are marginalized, engendering mistrust and disengagement from participants and risking disparities and disadvantage to individuals or groups. In this chapter, we discuss the importance of equality, diversity and inclusivity (EDI) in research and consider what support is available to work towards it. After defining the various terms that are used when discussing EDI, the chapter explores what they mean for researchers in different contexts. It outlines some initiatives such as Athena SWAN that are designed to promote gender equality and considers how these apply to research practice. The chapter will also detail what a researcher seeking to practise EDI in their work must consider at each stage of the research process, and it ends with a review of successful EDI interventions.

EDI: definitions

EDI is a global policy-driven initiative arising from the American Civil Rights Movement in the 1960s which set out to tackle racial discrimination. Initially, its aim was to increase the number of people from groups that had been disadvantaged in the workplace. Its focus has since expanded to mitigate discrimination against women, people with disabilities, those who belong to sexual minorities and many other groups (see below for list of Protected Characteristics). EDI policies have been adopted in many countries and a demonstration of their implementation is now a requirement of educational and other institutions. For researchers, this means not only familiarizing yourself with what EDI approaches mean for you and your research, but also to be cognizant of institutional expectations and support for incorporating it into your work.

Gilligan's ground-breaking study of theories of human development (1977) highlighted the exclusion of the 'feminine voice' in research and propelled the development of feminist perspectives. It has done much to raise awareness of the ways in which research can recognize and address discrimination and the propagation of taken-for-granted assumptions about women. Her work led to a greater focus on gender equality in societal influence and justice – a focus that continues to develop to this day. EDI in research now calls for not only gender but also class, age, race, ableism, sexuality and all other marginalized dimensions to be considered. Before we go on to consider how this can be done, let's first define what we mean by the various terms used.

- *Equality* calls for fairness and unbiased treatment of all individuals and groups. It seeks to ensure equal access to employment, study, healthcare, career opportunities and all other aspects of living a full life for everyone. One aspect of ensuring this as a researcher means considering inclusion and exclusion criteria for potential research participants.

- *Equity* is different to equality because it seeks to achieve equality by recognizing that advantages and barriers can be identity-based, and works to correct these imbalances rather than seeking to treat every person in the same way (Dewidar, Elmestekawy & Welch, 2022). Researchers in healthcare and education often work towards equity to ensure that interventions, diagnoses and treatments are meaningful to different individuals and communities.

- *Diversity* is the understanding of, respect towards and embracing of differences. It goes beyond acknowledging or tolerating differences to actively embracing them. Consideration of diversity should also include 'invisible' diversities such as people who are neurodiverse or who belong to a sexual minority. For researchers, including diversity means being proactive in seeking to understand and justify representation in their research, from team membership to participant recruitment.

- *Inclusion* means ensuring that all individuals and groups are part of and valued in an environment in which they can flourish (Xuan & Ocone, 2022). For researchers, this means taking steps to understand cultural, historical and other influences on groups and individuals so that data from them can be understood, as far as is possible, within those contexts.

- *Intersectionality* recognizes that individuals and groups have multiple identities because dimensions such as gender, race, class and so on intersect and reflect multiple systems of oppression (e.g. sexism and racism directed towards women of colour). Focusing on only one dimension obscures the unique experience of their overlap. For researchers, adopting an intersectional understanding often includes a commitment to social justice. It offers a useful approach with which to explore differences between and within groups but also presents challenges such as deciding which intersection to focus on or whether gender should always be included (Kelly, Dansereau, Sebring, Aubrecht, FitzGerald, Lee, Williams & Hamilton-Hinch, 2022).

- *Access/accessibility* is sometimes included in EDI policies (i.e. EDIA) to ensure explicitly the inclusion of people with disabilities.

- *D/Indigenous* is often included in countries with higher Indigenous populations (such as Canada and Australia) to ensure inclusion of people from Indigenous groups.

- *Protected characteristics* are characteristics on the basis of which it is illegal to discriminate against anyone. They differ across countries. For example, the UK Equality and Human Rights Commission defines protected characteristics as: age, disability, gender reassignment, religious belief, marriage and civil partnership, pregnancy and maternity, race, sex and sexual orientation (Equality and Human Rights Commission, 2021), and in South Africa, it is illegal to discriminate on the basis of race, colour, tribe, sex, language, religion, political opinion or affiliation, nationality, social origin, marital status, pregnancy, HIV status and disability (Xuan & Ocone, 2022).

European Commission Protected Characteristics (from commission. europa.eu)

Sex	Membership of a national minority
Race	Property
Colour	Birth
Ethnic or social origin	Disability
Genetic features	Age
Language	Sexual orientation
Religion or belief	Political or any other opinion

Understanding these terms and how they apply to your research is your responsibility and works to ensure that your research has relevance and impact. Incorporating EDI into your research practice requires consideration of these issues at every stage of your research. In the next section, we will consider how you can raise your own awareness of EDI in your research practice.

EDI and Research

We would like to think that most researchers agree with and seek to uphold the principles of EDI. Actively incorporating them into practice though can present some questions and challenges. Kelly et al. (2022) point out the increasing expectation of funders for EDI to be demonstrated in research and have called specifically for research into EDI itself. The expectation has led to a rise of 'health equity tourism'-research by researchers from disciplines outside the healthcare field. Kelly et al. point out that it is often the case that methods used by these researchers are the ones they use in other fields and cannot always incorporate the complexities and nuances appropriate to health research. They call for EDI training and support for researchers to start early in their careers to begin to minimize this. Whilst there are many ways for researchers in all fields to incorporate EDI principles into their work, and we will discuss these further in this section, the starting point for all researchers has to be their reflexive engagement with their own perspective on EDI and how they understand this.

Engaging with EDI

Researchers who are serious about bringing equality, diversity and inclusion into their research must reflect on their own social location, how this has been constructed and how it may enter the research. When considering EDI in, or as, your research, you must not only question what you understand about the terms themselves but also what they mean for you and the research. This requires an explicit self-consciousness and self-assessment about your own views and positions and how they might influence the research design and conduct (Booysen, Bendl & Pringle, 2018).

Often, this means being reflective about yourself and reflexive in how you are engaging with the research. There are many ways to do this, ranging from thinking, journalling and discussion to the use of conversational tools for use with team members (Purdy, Symon, Marks, Speirs & Brazil, 2023). The latter was designed to address the gap between academic conversations about EDI and questions about what to actually do when designing and conducting research. Designed as a conversation guide for simulation (sim) delivery research, the tool is used to ask questions such as:

- What aspects of this (sim) design, delivery and debriefing were related to gender, race, sexuality, culture, power, etc.?
- How did they unfold and with what impact?
- What are our potential biases or sources of privilege and power as they relate to sims today?
- Should we mitigate them differently, and are there individuals or groups that we should consult or involve?

It is easy to see how these and other questions that the tool prompts can be transferred to research in all topics.

But to return to you reflecting on yourself as a researcher, it is important to question implicit biases you may have and what these may mean for the research that you do. Bias refers to attitudes, behaviours and actions that are prejudiced in favour of or against others. Implicit bias is an unintentional and automatic bias that affects judgements, decisions and behaviour. Bias can be mitigated through recognition of biases you may have and training in strategies to reduce them. You can learn how people and groups can contradict stereotypes you may have about them and how to interrupt automatically biased thoughts to prevent implicit bias. You can ask yourself questions such as do I see myself as an insider or outsider to the participant group I wish to recruit from? What do I know about this group? How do I know this? How has the cultural and historical contexts of my own group influenced this? Armed with some honest answers to these questions, you can then develop plans and actions to gain other perspectives and insights. These might include considering where and how you access existing literature and research on the topic that you are planning to research, discussing your knowledge with others to gain different perspectives on it and challenging yourself to counter some of the justifications you have for holding these views. Whatever the outcome of this self-exploration and the questions you ask yourself, you will be equipped with deeper insight into how you see the world and, consequently, will be prepared to query and review your research conduct and decisions.

Positionality, reflexivity and EDI

Your researcher positionality is important when you are doing research that seeks to incorporate EDI principles. Understanding as much as you can about your positionality and how it intersects with issues such as power and privilege will help you to locate yourself in your research and better adhere to EDI principles. This means locating yourself regarding the topic of the research so that you acknowledge personal positions about it, locating yourself regarding the participants to consider how you view yourself in relation to them and locating yourself regarding the research context to acknowledge its influence on you and the research (Savin-Baden & Major, 2023). It is important to recognize too that you will have biases and subjectivities that you are not aware of, but by being reflexive and using this to understand more about your positionality, you will be better able to identify some of these.

Jessie King, an Indigenous academic who, in a positionality statement, identifies as Ts'msyen on his mother's side and mixed Irish and Scottish on his father's side reminds us that we need courage to pause and ask ourselves to acknowledge that, as researchers, we are exposed to multiple forms of knowledge and ways of knowing. Often though, academic systems insulate those that work in them from exploring barriers, structures and power dynamics that can cause us discomfort (King, 2023). Writing about personal experience as an academic, he proposes themes that can be used as a tool for self-exploration in relation to Indigenizing the academy. They are relevant for researchers in all topics.

Using self-reflection to lean into discomfort: King suggests that discomfort manifests from fear of the unknown, ignorance or feeling overwhelmed with where to start. He suggests four guide questions as a starting point:

- How much do I know?
- Where are the gaps in my knowledge?
- How do I fit into this conversation?

The fourth question specifically relates to Indigenizing the academy:

- Am I contributing to decolonization work that is supportive of Indigenous sovereignty?

The questions are designed to help understand 'What ignorance do I possess?'.

Promoting cultural safety: Employing practices based in cultural safety ('feeling free to embrace your cultural identity without fear of judgement, harassment, racism or discrimination' (Newton, 2021, pp. 7–8)) enables reflexive practice. In turn, reflexive practice enables you as a researcher not only to get to know yourself but also your relations with others. King advocates 'Etuaptmumk/Two-eyed seeing' in reflexive practice. This acknowledges realities we are living in and uses it to bring together strengths of diverse knowledge systems. The binocular effect capitalizes on the best of multiple ways of knowing. Incorporating respect, responsibility, relevance and reciprocity further informs your understanding and enables you to encourage and support relations with others.

Positionality, privilege and power: King argues that there is significance in knowing who you are (What makes you you?), and that it is important to make time to consider how this influences how you make decisions. It enhances understanding of how power and privilege intersect with positionality and enables grounding in authentic relationships with others. Recognizing the advantages conferred on individuals because of who they are, how they look and the positive perceptions of others can lead to feelings of guilt, anger and shame, but by understanding this, we can acknowledge what we possess and identify how it impacts our experiences, relationships and interactions with others.

Dancing with ignorance: King argues that acknowledging ignorance means we can work towards filling the gaps in our knowledge. Unpacking our ignorance can reveal a world of the unknown. Choosing to remain ignorant may, however, be founded in fear, perhaps of offending someone or making a mistake. King suggests being respectfully curious and being explicit when not knowing how to answer a question.

Working in ethical space: Ethical space is actively developed by people with diverse worldviews coming together with the aim of facilitating healthy, transparent and non-judgemental communication. They acknowledge different ontologies, epistemologies, cultures and communities, as well as the beliefs, values, stereotypes, biases and thoughts that comprise them.

This detailed and informative paper by King (2023) acknowledges that exploring these themes to challenge ourselves and the relations we have with others can be uncomfortable. It emphasizes the need to consider the philosophies, usually with Western-based perspectives, that exclude Indigenous values and dominate the structures and systems that define our concept of society. In doing this, we can better understand how they are incompatible with Indigenous ways of knowing and being. It emphasizes that ensuring that the responsibility for making space for multiple ways of knowing and being in relations is shared can promote development of ethical spaces and potentially lead to the transformation of colonial spaces through indigenization and decolonization.

There is much that can be taken from this paper to understand both the personal challenges to understanding ourselves and what we bring to the research, and the confronting and addressing of ways of better promoting equality, diversity and inclusivity in research that ranges across many groups.

Positionality statement (from King, 2023)

Hadiksm Gaax di waayu, Jessie King di waayu. My name is Jessie King and my traditional name given to me in 2008 is Swimming Raven. This name was given to me by the head of my family at the time, bestowed upon me by family members from Gitxaala where I trace my matriline. I am both Ts'mysen on my Mother's side and mixed-Irish and Scottish settler on my Father's side. I honour both of these identities, as co-existing, and I walk each day sharing this positionality out of respect and acknowledgement of what inherent privileges, power, and positioning this gives me in my life. I grew up with privilege, I maintain a level of privilege simply because of who I am, where I come from, and what experiences I carry with me. I did not grow up in spaces where I endured the racism and discrimination many of my family members and friends did. I did grow up not knowing of colonial history or hearing the language of my Dzi'is (Grandmother), I am doing that learning now. It is a challenging conversation to have but essential to share as this is the first gift you can give someone. It is also with a mindfulness of how I read research, journals, and books with Indigenous content – I want to know who the author is. This is who I am.

EDI in Research Practice

Striving towards EDI as a researcher means being aware and seeking to practise it throughout the research process. In this section, we consider some questions we can ask of the research design and conduct that can help with this.

The literature review

Reviewing existing literature about the topic you are proposing to research helps you to understand what is already known, think critically about what it tells you and identify a gap in the research that your study might start to fill. From an EDI perspective, useful questions include:

- How diverse are the sources of the literature you review?
- Do they include a range of philosophies, studies and perspectives?
- Have you brough bias to the literature you choose to review?
- How do you critique literature with a different philosophy of the world to your own?

Research question/hypothesis

The research question/hypothesis is generated from the literature review to inform the focus of the research. From an EDI perspective, you should ask:

* Does it incorporate biases, assumptions or taken-for-granted knowledge?
* How can these be recognized and/or eliminated?
* What limits will the research question/hypothesis put on the research and participants in it?

Participants

Participants from whom data can be collected are crucial, and EDI researchers aim to ensure that they can be reached and engaged with in the most accessible ways possible. This means asking questions such as:

* What are the implications of the inclusion/exclusion criteria?
* How can participants be stakeholders and collaborators in the research?
* How do the ways of attracting participants to the research include/exclude individuals and groups?
* How am I ensuring meaningful consent and debriefing?

Methods

The methods for eliciting and gathering data aim to collect information most pertinent to the research focus. EDI researchers select methods that strive to ensure that a wide range of participants who fit the research criteria are able to contribute. This means asking yourself questions such as:

* Is the language I am using to elicit the data inclusive and accessible?
* Are forms of data elicitation other than verbal more appropriate to ensure diversity and inclusion?
* Are the sites of data collection appropriate and accessible?
* For online data collection, have I used appropriate colour contrasts, sound levels and explanations of symbols and words?

Data analysis

All data analysis will be systematic but is also inclusive of subjective decisions. This can mean implicit bias and assumptions can enter into it. EDI researchers should ask themselves:

* Why have I chosen this method of analysis?
* How will I treat results that do not seem to fit or are outliers?

- How have I considered participants' choices of response?
- What is the analysis provoking in me?

Writing up

Writing up research means presenting it to an audience for interest, assessment or review. Whilst there are expected formats for different forms of writing up, the content and how it is constructed are down to the researcher/author. EDI researchers should ask themselves questions such as:

- Who will this write-up be accessible to? Who will it exclude?
- What biases have I included in the reporting of the research outcomes?
- How has my pre-existing knowledge informed the write-up?

Dissemination

Effective dissemination of research means reaching the audiences it is intended for and ensuring it is accessible to them. EDI researchers will ask themselves questions such as:

- Who will this form of dissemination exclude?
- How can I be more transparent about the research and its meaning?
- What are different ways of disseminating the research so that it is meaningful to more people?

Using questions such as these, and combining them with the themes outlined in the section above, will help with developing EDI practice as a researcher. They will also aid in identifying inherent assumptions you have brought to your understanding and practice of research. Discussing them with others can be useful in exploring challenges they present to you as a researcher and in your researcher role.

Reflective question

In what ways do you (unintentionally) exclude potential participants or audiences for your research? What role do your own biography and experience play in this?

So far in this chapter, we have considered what it can be like for researchers wanting to develop EDI research practices and how they can work towards this. Whether you are working alone or as part of a team, the promotion and practice of EDI in research necessitate self-reflection and understanding of your positionality. This can be supported by institutional structures and services, and we discuss in the next section how these are provided.

Support for EDI researchers

EDI policy initiatives originated from Acts initially developed to address gender and race discrimination in the workplace. Employment Equality Acts outlaw discrimination in work-related areas such as pay, access to employment and promotion prospects. Lists of Protected Characteristics detail the characteristics of individuals on the basis of which illegal discrimination can be claimed. The aims of such Acts are to eliminate discrimination in the workplace to create work environments in which every individual feels they belong. In universities and other research institutions, EDI initiatives range from training opportunities to individual support to embedded charters designed to support and transform EDI practice.

There are many reasons why EDI initiatives for university researchers are beneficial. They offer opportunities to train researchers to incorporate EDI into their work and also encourage EDI in who and how people are recruited to work within, or in affiliation with, the university or institution.

ECRs are future research leaders and, so, should be fully involved from the outset in decision-making about, and shaping of, research structures. Diversity amongst this group brings multiple perspectives to systems that may have until recently been developed by senior researchers who may, in turn, have been part of recruitment and promotion approaches that did not encourage diversity in ableism, racism, ageism, etc. (e.g. Clark & Hurd, 2020). Now, universities are actively seeking to recruit from wider ranges of age, gender, sexuality, race, ethnicity, disability, geography and so on. This not only enables more creative and generalizable solutions and developments in research but also actively includes input from people with direct experience of living with these characteristics. Arguably, ECRs are more likely to be open to new solutions and the use of new research tools and techniques, and this brings a freshness and new opportunities for EDI in research from the group that often makes up the majority of the research community.

There are also benefits to supporting experienced researchers in EDI initiatives. These can range from offering advice and training in new EDI terminology (for example, referring to ethnic minority groups individually rather than using BAME as an umbrella term) to providing training in methodologies that may be better suited to EDI research (such as Participatory Action Research) than the conventional methods they may be more used to and familiar with.

A specific policy for encouraging gender equality in university environments in the UK, Ireland, Australia, Canada and the USA is Athena SWAN (the EU Commission is currently exploring the development of a similar Europe-wide scheme) (Kalpazidou, Schmidt, Ovseiko, Henderson, et al., 2020).

The Athena SWAN Charter was established in 2005 to support and transform gender equality within higher education (HE) and research. It is a framework that encourages and recognizes commitment to advancing the careers of women in science, technology, engineering, maths and medicine (STEMM) employment. It is now being used across the globe to address gender equality more broadly, not just barriers to progression that affect women (advance-he. ac.uk). It consists of three levels, Gold, Silver and Bronze, that recognize the efforts that universities make to promote gender equality, and the Charter is awarded as a result of data collection and monitoring by universities to demonstrate their success.

The Athena SWAN Charter (from advance-he.ac.uk)

- Helps institutions to achieve their gender equality objectives.
- Assists institutions to meet equality legislation requirements, as well as the requirements and expectations of some funders and research councils.
- Uses a targeted self-assessment framework to support applicants in identifying areas for positive action as well as recognize and share good practice.
- Supports the promotion of inclusive working practices that can increase the retention of valued academics and professional and support staff, demonstrating an institution's commitment to an equitable working environment.

The Charter has EDI as a core principle and this applies to the recruitment and treatment of both students, and teaching and research staff. The Charter and the universities who adopt it commit to practices that promote seeking out individuals from marginalized groups for employment by them, removing barriers to participation (such as inaccessible language or materials) and learning about the impact of implicit and explicit biases (such as disparities in research funding grant allocation). In order for it to be successful, institutions may need to be prepared to allocate extra time, money and other resources. One such resource is the provision of EDI mentors for researchers.

Fostering EDI

Mentors

Mentors are different to research supervisors or advisors because their role is not focused on the research as such but on you as the researcher, with the aim of supporting and facilitating your development as a researcher. The mentor-mentee relationship is a flexible and reciprocal one in which an experienced researcher shares their knowledge with you to help progress and achieve goals that you set together. You learn from one another and craft the relationship in accordance with your needs and skills. This can be very helpful if you want to develop ways of incorporating EDI principles into your research and, indeed, if you feel that the group you belong to is under-represented or marginalized.

Mentors can help increase the diversity of your research by helping you to access their networks for collaboration or recruitment. As an experienced researcher, it is likely that your mentor has a wide range of contacts and colleagues who may be interested in hearing about and supporting your research. They may also have ideas about why and how previous research has not ensured inclusivity and diversity in it, so they can help you to think about how to avoid this in your future work. They are also likely to know what support the university has for EDI and who you can talk to about issues with your research and well-being. Using your mentor as a sounding board for concerns with the limitations of your research can lead to creative solutions to recruitment challenges. In turn, you or your mentor may have knowledge of a method that is more appropriate to including EDI in the topic that you are researching, and you may be interested in learning about it together. An important aspect of the mentor role is to help with your career development. As the importance of EDI is increasingly recognized, the mentor's help with instituting it into your research has the added benefit of helping you towards goal-setting and promotion.

It is not always the case that universities, research institutes and other institutions in which you may be working as a researcher have a formal mentoring scheme in place, but there are resources available that can help you find a mentor. One of these is AuthorAID (www.authoraid.info). AuthorAID aims to help researchers in developing countries to write and publish their work through access to a global network of other researchers. Members can search the network for researchers available to offer support and advice on aspects of their research and then make contact with them. The network covers a range of disciplines.

In the absence of a formal mentoring scheme in your institution, you can always approach a more senior academic colleague and request that they act as your mentor. This may be someone you have worked with as a research assistant and who you know has an interest in you and your career progression. Similarly, it is not unusual for PhD supervisors to become mentors to PhD candidates after the doctoral studies have been successfully completed. The focus is frequently on learning how to get published, and next, we will consider how academic journals are striving to improve equality, diversity and inclusion in research.

Publishing research

As a researcher, there will be expectations of how you conduct your research, and you will have your own aspirations of the researcher roles you want to take up during your career. In academic institutions, this will almost certainly involve publishing research papers in academic journals. You may also be asked to review papers written by other researchers and to join editorial boards as your career progresses. There are an increasing number of strategies that you can employ to promote and incorporate EDI in these tasks.

As an author, researcher or reviewer, you are a stakeholder in and gate-keeper of rigorous and ethical research practice and dissemination. You may have to make critical judgements about what to include/exclude from the paper, and you have a responsibility to ensure that you take steps to acknowledge and address bias (intended or unintended) and minimize its perpetuation (Dewidar, Elmestekawy & Welch, 2022). The academic publication process is built on objectivity, gender and socio-cultural neutrality, and respect for the rights of humans and animals (e.g. Baveye, 2021; Kaatz, Gutierrez & Carnes, 2014). To address these, journals are increasingly including an EDI statement about what is expected in manuscripts submitted to them. These may be a definition of the problem of lack of EDI so that a shared understanding can be acknowledged and addressed by all stakeholders in the publishing process, requests that information about the representativeness of the participant sample be included and clarity about the bias-free language and style that they expect from authors (e.g. 'they' as a pronoun rather than assumptions about 'he' or 'she'). Some journals have an EDI lead who actively works to advocate and promote EDI by, for example, developing recruitment strategies to increase diversity on editorial teams. Journal editors and EDI leads may also take steps to advocate for increased awareness and practice of EDI principles by, for example, providing training for editorial team members and developing advisory committees that include academics and other researchers from under-represented communities. Diversifying advisory and editorial teams will bring multiple perspectives to their discussions, increasing the likelihood of cultural competency in decision-making and the development of practical solutions to EDI awareness and practice. Monitoring EDI data about the journal and the publication process and using their position of power to advocate for EDI means that journal editors and others involved can play an important part in cultivating EDI.

EDI researchers as reviewers

Researchers are often asked to review manuscripts submitted to journals for consideration for publication. You will be approached because you are considered knowledgeable in your field and also in the standards and presentation of research. The role requires you to consider the research, its quality and how it fits with the aims and scope of the journal. Holding this in mind means that you will comment on strengths and limitations of the manuscript and proffer a decision on its acceptance, amendments needed or rejection. As a reviewer who holds to EDI principles, you should consider how you write your review so that it is clear, constructive, respectful and written in an inclusive language. You can also, however, consider whether written feedback is the most appropriate way – perhaps people with neurodiversity would appreciate a phone call as well to offer extra feedback, for example. It is useful to hold in mind too that that the author of the manuscript may not have English as their first language, and this may be apparent in the writing style of the manuscript. Rather than rejecting a manuscript because it does not meet the required writing standard of the journal, you can consider suggesting resources for support in rewriting it for resubmission. SAGE and Elsevier, who manage several journals, both offer language support services for authors.

If as a researcher you enjoy reviewing manuscripts, you may also be interested in joining review boards and editorial teams. Journals are starting to offer mentoring schemes that encourage working towards these leadership goals, and here, you can learn (and teach) more about how EDI can work in practice in the publishing world.

It will also be the case that if you are a member of an under-represented group, you are in a strong position to highlight your usefulness as someone who can bring your perspective to these roles. Similarly, you can draw on your community network to encourage and recommend others to take up similar roles.

Using the strategies and initiatives that journals are taking and applying them to yourself as a researcher will help the field of publishing to continue working towards recognizing how bias or prejudice can be perpetuated. You can be an active contributor to finding ways to challenge a lack of equality, diversity and inclusion by raising awareness, advocating and contributing your learning, experience and expertise in a number of constructive ways. This should mean that creative solutions with achievable action plans can be developed and progressed, and one day, you may be one of the leaders making this happen.

EDI researchers as supervisors

As your research experience develops, you may be asked to become a research supervisor. In this role, you may work as the sole supervisor or one of two or three to support and guide a student researcher through to qualification. This role is very important for researchers who are starting out because it will likely be held over a number of years with the student and will have influence over the researcher they become as well as over their research practice. It also provides a vital platform from which you can support and inform students of the importance of EDI – for them as individuals, for their inclusion in the researcher community and for their approach to research.

The supervisor–supervisee relationship has inherent power dynamics within it, and as a supervisor, these should be recognized and appropriately navigated. Supervisees are not only dependent on you for their progression through their researcher training, but also for receiving funding and being supported to publish their work. The relationship will also extend to you providing references to future employers and considering the supervisee for further research opportunities. In addition, supervisees are likely to develop new identities as researchers and will look to you for guidance on this.

These factors place supervisors in powerful positions, however good the relationship with the supervisee may be, and mean that you as a supervisor have a responsibility to practise EDI within them. Knowing that you are likely to have implicit biases about an individual's characteristics or abilities helps to find ways to recognize and address them. Not doing so can exacerbate the power dynamics to the likely detriment of the supervisee's work and sense of self. In addition to being reflective about implicit biases you may have, it can help to take Implicit Bias Tests. Harvard University has developed a series of tests that are freely available online and help you to recognize bias and any associations you may have in a range of areas such as age, gender, weight, sexuality and disability. There are also tests for associations with cultures, race and beliefs (see implicit.harvard.edu for the full range of tests).

Knowing what biases you may have can help with consideration about how you express values about research and researcher characteristics to your supervisee. This is important because as supervisees learn to become researchers, they are likely to absorb your expressed values as guides to forming their researcher identity. Undermining the supervisee with generalized or incorrect assumptions about their abilities because of their membership of particular groups or communities risks excluding them from feeling that they belong to the researcher community (as well as being insulting and disrespectful). Pay attention to ensuring that you do not disempower or offend supervisees with the language you use, and be open in telling them you are interested to understand how best they can learn from you. Knowing the needs, expectations and values placed on the supervisory relationship by the supervisee is more likely to lead to their productive development as a researcher.

Finally, don't underestimate the importance of being a role model to supervisees. They will see you as an experienced and respected researcher, and how you conduct your research and guide them in theirs is likely to have a lasting impact on their development and, perhaps, their future role as a supervisor themselves.

EDI researchers as teachers

As an academic, practice-based or community-based researcher, you may well be involved in teaching and training. Whether this is as an assistant as your research career takes off or as an experienced researcher who is expected to take up this role, there are many opportunities to incorporate EDI awareness and principles.

It is key to remember, and emphasize, that students learn differently, regardless of their abilities. Enabling and supporting diversity in student groups mean finding ways to ensure that every student is facilitated to reach the scholarly standards expected. As a teacher, it is always good practice to find different ways of presenting the same information. This can mean using visual presentations as well as oral ones, summarizing as you go along and encouraging group and individual work. Extending this to testing and assessing students means that different formats should be available – poster and verbal presentations, test completion at home and, if needed, allowance of extra time. It should be made clear to students that if they need interpreters, note-takers or other forms of support, this can be accommodated, and they should be informed of all the institutional support systems available. Many teaching environments now include closed caption and lecture recordings and make materials available in large format or braille. Institutions also have departments dedicated to providing practical support, counselling and technological aids to remove barriers to learning.

It is also important to remember that lateness or periods of absence may be explainable by challenges such as navigating crowded corridors in wheelchairs or experiencing episodes of mental ill-health. The best way of raising your awareness of students' needs may be ensuring that you convey that you are interested and wanting to learn about them. Developing an open and welcoming style helps students to feel comfortable to share them with you or, at least, with the institutional support structures who can bring it to your attention. An openness to awareness of the impact of disability and other issues that can challenge students' inclusion can be conveyed by explaining what steps you are taking with the group in general, for example, ensuring that you are speaking to the class in a way that lip readers can follow. With this openness, it may be that individual students feel able to approach you and explain what they need. As an EDI researcher who also teaches, you can then make changes in your teaching style and environment.

Much teaching involves written materials such as papers and textbooks. Consider carefully what your recommended reading lists comprise – is there an inclusion of authors from countries other than the one you are teaching in? What languages are they available in? Do you make them available in advance of lectures to enable extra time for them to be read? Are there visual materials such as graphic displays or mind maps that can convey information in ways other than linguistically? Also, draw on your own research to demonstrate how EDI can be incorporated and learnt about. Many students like to hear about lecturers' and trainers' experiences of research, and this can be a useful and beneficial learning hook to keep student interest.

As an EDI researcher who teaches, it is also important to be aware of issues such as depression or anxiety amongst students. They may or may not want to talk to you about it but it is important to recognize the limits of your role and to make yourself aware of appropriate support services that may be available to these students. These can be shared with the student group so that they can avail of them if they wish. Similarly, if a student's behaviour in the lecture or training session is disruptive or threatening, remember it is your role to discuss only their behaviour with them and the expectations of conduct in the learning environment, not to label or predict the reasons for it. Remember too that anxiety can worsen speech and language for some, so an openness to reasons for a lack of communication is also beneficial.

To summarize, there are a number of practical supports that can be incorporated in EDI teaching environments but the ways in which you present yourself as the teacher can give additional insight into how best to support diverse and inclusive student groups. For you as a EDI researcher who teaches and trains, this means leading by example, being aware of how you convey a sincerity about wanting to learn about student needs and developing relationships with students that enable them to discuss and explain what would help them to reach the standards expected.

Chapter summary

In this chapter, we have discussed equality, diversity and inclusion from different perspectives. Whether this is as a researcher carrying out research, reviewing other people's research or teaching people to do research, the importance of knowing ourselves, how our positionality can have an impact and how we address implicit bias has been emphasized. This is not only to take up our responsibilities to promote and advocate for EDI in the researcher community, but also to do what we can to ensure that future researchers will feel valued and inclined to advance EDI practice.

Further reading

Grace, S., & Gravestock, P. (2008). *Inclusion and diversity: Meeting the needs of all students (key guides for effective teaching in higher education)*. Routledge.

This book highlights good practice for all students and provides a helpful structure around the day-to-day experiences of staff and students as they make contact with each other. Whilst referencing international literature, and discussing some of the educational principles that underpin an inclusive curriculum, it offers guidelines on different aspects of teaching and learning.

Malone, T. (2021) *Equality, diversity & inclusion: The practical guide: The essential handbook for terminology and communicating inclusion with dignity* (2nd ed.). Independently published.

This book provides a practical guide for current EDI terminology and focuses on maintaining dignity and respect for communities. This updated version includes key terminology in the wake of COVID-19 and the Black Lives Matter movement.

Thomson, A. (2024). *Nurturing equality, diversity, and inclusion: Support for research careers in health and biomedicine*. Policy Press.

This book familiarizes readers with key EDI issues in relation to research careers and researcher development. It uses an evidence-based approach to bring together challenges and solutions to EDI matters and offers practical strategies and interventions for academic and research settings.

Part 4

Researcher Identities

10 Being a Researcher: Theory and Practice

Introduction

The premise of this book is that being a researcher is both something you learn and someone you are. If you are investigating a topic for assessment, professional practice or with a community, it is likely that you will receive training and teaching about what the research process is and techniques for carrying it out. We have seen, however, that there is also a very human aspect to doing research, whether it is simply your enjoyment of it, a self-reflective awareness of your role as a researcher or your positionality about the research. In this chapter, we look at both what you are taught and told about being a successful researcher and what it is like for you to be a researcher. We will start by looking at what and how research practice is taught, before moving on to discuss how you as a person understand and develop yourself as a researcher. This will incorporate considerations of power dynamics, being and feeling like an insider or outsider researcher and some ideas about why and how you take up positions during the research or are positioned by others. We will think about how you recognize and address these issues and what they mean for your enjoyment and success as a researcher.

What you are taught

Depending on the environment in which you are being taught about becoming a researcher, you will receive training in approaches to research, techniques of research, the researcher role and impact in research and how to disseminate research outputs. We have seen in Chapters 4, 5 and 6, some of the pathways to qualification and practice as a researcher. Following qualification, some choose to progress their research career by becoming a recognized researcher. There are various labels and lengths of time for this role (see box below) but all reflect the growing recognition in Europe and beyond of the importance of researchers (Eley, Wellington, Pitts & Biggs, 2012).

Different definitions of 'researchers' (adapted from Eley, Wellington, Pitts & Biggs, 2012)

Early stage researcher: Salzburg Principle No. 4, Europe not including UK, first four years of PhD student research experience or until doctoral degree awarded, whichever is shorter

Early career researcher: UK Arts and Humanities Research Council, within eight years of PhD award or equivalent experience or within six years of first academic appointment

Experienced researchers: European Commission, at least four years research experience since gaining a university diploma in the country in which the diploma was obtained, or researchers with a doctoral degree (no time frame)

Early Career Investigator/Early Stage Investigator (National Institute of Health, USA): within 10 years of highest degree in an academic field or professional discipline, or medical residency

It can be seen that the length of experience that defines the beginning of a researcher's career varies with continent and country, from doctoral student to up to 10 years postdoctoral research experience. These definitions can be seen as most useful when applying for jobs or research funding. What is of more interest is who these people are and how they develop their researcher persona. What we can take perhaps from this range of definitions is the value of experience, and that is particularly interesting when we think about how formal teaching of being a researcher tends to end at doctoral level, at the latest. From then on, it becomes more about mentorship, relationships with advisors and team leaders and self-awareness. We have discussed these in previous chapters, so in the next section, we begin to delve into what you teach yourself as you develop into a researcher.

What you teach yourself

An early lesson that researchers learn is the importance to them of the topic they choose. Whatever the stage of your practice as a researcher, the research will take a long time or feel like it is taking a long time. People new to research, carrying it out because it is part of an academic or professional training, are often juggling the research with other assessments and practice. Doctoral researchers can expect to spend a minimum of three years on their research, and postdoctoral and beyond researchers can spend any number of years on a study or series of studies. Researching a topic that you have an interest in is probably a minimum requirement, and researching a topic you feel excited and passionate about will actively help you to persevere with it and likely keep your motivation at a level of enjoyment rather than duty.

Researchers are passionate about the topics they research for a number of reasons: you may have a burning interest arising from personal or professional experience, you may want to be at the forefront of emerging research, you may have political or cultural interests that mean you want to address injustices or you may want to continue research into a topic you first started researching earlier in your career. Being excited about your topic and the research into it will help to maximize your dedication to it and will also keep you going when you encounter challenges, frustrations and delays in it. It will also help you to think creatively when you reach results you were not expecting (or wanting) so that you are accepting of them and see the value of the research and why these results were reached. Although you may still feel disappointment, your researcher stance will mean that you can accept this and move forward – by reporting the results with explanations of why they were reached and by re-developing the research to test these against a new way of examining the same problem. If you are confident that that it has been carried out rigorously, this is preferable to abandoning or burying the research

Excitement

Choosing a topic that excites you can be the easy part – for you to maintain your interest in it can depend on how and what you ask and test about it, how worthwhile it is to a wider audience, how feasible the research is within the timeframe you have to work to (including accommodating time for your other tasks and practices) and who the other researchers supporting and mentoring you might be. However, it may also be that there are times when you have a limited choice about what you research. A supervisor may nominate a topic or you may want to work in a particular institution or with a particular community whose members have a clear idea about what research is of value to them. You may need funding to maintain and pursue your career and, therefore, have to tailor your research to areas likely to attract this. There is still scope for excitement about a topic even if it is not your first choice – relating some aspects of it to aspects that interest you about your number one choice can awaken that excitement (exploring why mothers attend parenting classes rather than exploring why mothers choose to return to work, for example). Excitement about a topic that is not your first choice can also develop because it gives you an opportunity to gain new insights into it, see it from different perspectives or learn new methods of research that you can plan to use later when you are able to explore your topic of choice.

Relationships

As we have seen in other chapters in this book, even if you are the sole researcher on a study, the support of trusted, experienced others with whom you have a good relationship can be invaluable both in celebrating the mini-achievements of your research as it progresses and in addressing challenges as they arise. This emphasizes again the importance of investing time and effort in your researcher relationships. Is there someone who inspired your interest in a topic by the way they talked about it (teaching or in their own research, for example) who might be a supportive supervisor or mentor to you in developing new research in the topic? And is this person someone who you can see yourself having a productive and, perhaps, enjoyable, working relationship with? The power dynamics inherent in supervisor–supervisee/mentor–mentee relationships can be harnessed to your benefit if you engage with them. This means acknowledging the superiority of the supervisor/mentor in research practice and expertise but also being willing to share with them the knowledge and skills you are developing. This can be an enjoyable process for both of you and add to the excitement you feel about doing the research. Similarly, discussing and debating issues arising in the research with a supervisor/mentor can lead you both to new insights and understandings, again giving a new angle to being a researcher.

For this relationship to be successful, however, you have to be the sort of person who is willing to be open and communicative, someone who wants to learn from their supervisor/mentor and not simply have them as a requirement. There may be times when they can feel like they are criticizing you and this can be upsetting. Are you the sort of person who can see criticism as being about the research and not about the researcher or will you take everything to heart and feel like a failure?

Apart from the practicality of setting clear guidelines (or contracts) with your supervisor/mentor so that you both know what is expected of you and of each other, this can also help to disperse feelings that they are not interested in you or your work. You may have to curb the expression of your excitement to or the seeking of answers from them about problems in the research until they have available time to give you. Having established early on how busy they are with their own research or other tasks will help your emotional responses to be more manageable. When you are immersed in your research, and have a good relationship with your supervisor/mentor, it can be easier to understand and accept that they have similar commitments with other researchers they may be mentoring or supervising.

Knowing whether you are a person who prefers to work alone or as part of a team can help you to learn how to develop productive and enjoyable relationships with other researchers. Again, clear guidelines can help – are you expected to respond to or send midnight emails to a team leader or other team members, for example? – and you can learn how to overcome feelings of frustration or resentment as and when they arise. Are you someone who listens well? There may be team meetings when other members are discussing aspects of the research which you do not understand or talking about a detail you think unimportant. Rather than being the person who interrupts or puts down other people, you can learn to hear them out and raise appropriate constructive questions so that the whole team has a shared understanding of each other's expertise and how it is used in the research. In other words, we can add patience and engagement to openness as researcher characteristics that can benefit both you and the research. Similarly, when you are describing your input to the research, are you doing it in an accessible manner? Often, researchers are so engrossed in what they are doing in the research that they forget that others have not come on the same journey with them. Your own passion for a topic and its research may not be matched by others, who may have different agendas and motivations for working alongside you on it. The researcher who can explain their role and work in a study, and respond to questions in accessible ways, is more likely to enjoy team meetings as well as the research that they are doing.

Family and friends

In a different context, much of what we have said above can apply to your relationships with your family and friends. However much you enjoy being a researcher, it is not the only person you are nor the only thing you do in your life. Your family and friends may have to accept that you are absent, physically or mentally, for periods of time, and if you can understand what this might mean for them, you can offer reassurance and information about how long it is likely to be for and negotiate time together that you are all happy with. This takes a researcher who can look at the impact of their role from many perspectives – using expertise and learning skills for research, conveying the research to other researchers and to lay audiences (who may be interested family or friends) and recognizing that there may be times when you have to temporarily walk away from the research to invest time in the preservation of other relationships. You may well want to draw support from these as the research hits bumps or achievements, and it is often in the quieter times that care can be given to others around you so that mutual interest in the research, and the person carrying it out, can grow.

Anxiety

Many researchers describe experiencing anxiety as they approach and carry out research. Although most often discussed and documented in relation to doctoral researchers, it has also been described in academic and professional researchers at all levels of their careers. Anxiety may 'flare up' or be a constant background for individuals. It is not hard to see how and why it can 'stick' to research: the range of roles that a researcher has to take up, the different research spaces that the researcher has to enter, the interactions with participants and other researchers, supervision meetings, writing up for different audiences, verbal presentations, timeframes, concern that the research will be in vain or completed before you by someone else and so on can all trigger or exacerbate anxiety. Anxiety often manifests as an embodied experience (sweating, dry mouth, rushing blood) provoked by thoughts of self-doubt, sense of failure, feeling like a fraud and guilt. In the research environment, if it intersects with feelings of being othered or not belonging (Edwards, Holmes & Sowa, 2019), it is likely not only to influence the research but also your sense of self as a researcher. Recognizing this and harnessing it can, however, be of benefit to your researcher role. Self-reflection about your experience and manifestation of anxiety and learning about what aspects of the research or the research environment may trigger it can help you to see how and where it might influence your research practice. For example, if you are a shy person in wider social contexts, you may be particularly uncomfortable at the prospect of approaching potential participants (see e.g. Scott, Hinton-Smith, Härmä & Broome, 2012), so you may design your recruitment processes differently, such as by using online communication. If you are someone who prefers to work alone, the prospect of research team meetings may be anxiety-provoking and you may avoid attendance or contribute less to the discussions. On a broader level, the intensity of experiencing anxiety can infiltrate your positionality and subjectivity in research so that you are concerned about who you are as a researcher, how you are perceived and what it means for the research.

On the other hand, there are some very real reasons to feel anxious – what if someone else really is carrying out research in the same area as you and is likely to conclude it before you do? – but when the anxiety becomes overwhelming, you can be at risk of becoming stressed or burnt out and withdrawing or halting your research.

There are ways, however, in which you can teach yourself to navigate and manage anxiety as a researcher.

You are likely to have aspirations to present as a good researcher. The dominant discourse of what a researcher is will have been formed by what you have been told (see discussion on being a supervisor in Chapter 9), what you read and the perceptions of other researchers you know. Until recently, researchers have been predominantly portrayed as white, male, from higher socio-economic backgrounds and with conventional formal education. When you look at research leaders, they too are likely to fit this ideology. Increasingly, however, researchers in academic, practice-based and community settings are showing greater diversity. Rather than changing the meta-narrative, however, this may lead to a greater sense of being an imposter or feeling you have to respond to stereotype threat, believing that other researchers are not like you.

Writing or journalling can help you to understand what this means for you. Edwards (2019) and Todd (2021) suggest using writing practice to interrogate and consider the different dimensions of who you are and what they mean in relation to who a researcher is. In your writing, you can ask what failure in your research means – what will happen if the research does not go as you thought it would? What can you learn from the apparent failure and how can it shape the next steps in your research? It is usually 'successful' research that is publicized, but you, and others, can learn from mistakes and missteps (Gaillard, van Viegen, Veldsman, Stefan, & Cheplygina, 2022). Stefan (2010), Professor of Physiology at Medical School, Berlin, and head of a computational neurobiology laboratory, who has held several prestigious fellowships and has many publications, suggests keeping a record of your failed applications and research to compile into an alternative CV that demonstrates not only your full research history but also how you addressed them to write further grant applications, papers and research proposals. She highlights that researcher narratives are constructed around success so that other researchers seem to have a constant stream of research triumphs. We are not told of the research that was rejected or failed, nor of the many hours of work that led to this. When we experience failure as a researcher, it can feel like we are the only person to do so, and this can lead to isolation, dejection and anxiety. Making the invisible visible in this way can enable you to see your perseverance, the lessons learnt and the benefit to the successful research that you have done. Publicizing it, if you dare, can inspire and help other researchers.

Imposter syndrome

Imposter syndrome (Clance & Imes, 1978) was first coined in the 1970s following research with women in academia. The researchers found that women frequently felt like frauds, unworthy of the praise they were given for their academic or professional achievements and that their talents were overestimated by others. It is now known that all genders are affected equally by imposter syndrome, despite it being most commonly associated with women (Mann, 2019). Clance and Imes (1978) found that women work hard to prevent others from discovering that they are an 'imposter' by paying diligent attention to concealing their true opinions and ideas so as to only voice those they think will be well-received (intellectual inauthenticity). Additionally, they seek to gain approval of their superiors so that they will be well-liked and perceived as intellectually special. With the belief that society in general rejects successful women, they also avoid displays of confidence (Edwards, 2019).

Imposter syndrome can be linked to 'stereotype threat' (Steele & Aronson, 1995), in which people from marginalized groups feel at risk of confirming negative stereotypes of group membership and actively seek to contradict these perceptions. The interpretation and internalization of how you are perceived by others can underscore the production of anxiety as people feel they do not belong or must prove they do belong in the environment they are in (Edwards, 2019). Context is clearly of primary importance in the perpetuation of these positions and behaviours because looking to others to see the characteristics of authentic researchers (or practitioners or professionals) means that, often, you notice, and focus on, differences, which, in turn, make you feel like a fake. Of course, we cannot know fully how others perceive us, so the adjustment in behaviours to try and fit in are often formed from assumed perceptions. In turn, these may be correct or not, so rewards are not always guaranteed... leading to more anxiety.

To address this, you can question comparisons between you and the portrayal of other researchers. Are there successful historical researchers who do not fit the common portrayal of a researcher? Think of Marie Curie, Rosalind Franklin or Stephen Hawking, for example. George Washington Carver (1860s to 1941) was an American born into slavery and became a botanist, inventor and teacher who invented over 300 uses for the peanut and developed methods to prevent soil depletion. He advised Mahatma Gandhi and President Roosevelt on agriculture and nutrition and was made a member of the British Royal Society of Arts – a rare honour for an American citizen.

Increasingly, contemporary researchers also include figures who do not fit the dominant profile of a successful researcher. Kimberlé Crenshaw, who is the Isidor and Seville Sulzbacher Professor of Law at Columbia Law School and a Distinguished Professor of Law at the University of California, pioneered critical race theory and proposed the double bind of gender and race prejudice (intersectionality) (e.g. Crenshaw, 1995), which serves to perpetuate oppression and inequality. Her work has pioneered the development of feminist research around the world. Temple Grandin is a Professor at Colorado State and an expert in animal behaviour. She describes herself as an autistic woman and wrote the first text giving an inside view of living with autism (Grandin, 1992). She describes the frustration of not being able to speak in her early years and how early intervention enabled her to go on to develop an academic career in which she is widely lauded. By familiarizing yourself with the many figures who are members of marginalized groups and have become leading experts in research, you can look to understand why and how oppression in academic and other systems may be holding you back and how others have overcome it.

On a similar note, it may be that your quest for support is within a system that appears not to support members of your group or with which you may not feel comfortable. Perhaps, available supervisors, mentors and advisors have different characteristics to you or have found that their positions of privilege and power preclude them from changing their beliefs and behaviours. It can be helpful then to look outside the research environment in which you are based to other groups of people who experience the same. This might be groups (online or otherwise) that are made up of researchers from the same group(s) as you, groups of researchers who hold similar beliefs as you (e.g. religious groups) or groups of people at the same stage of their research career as you (PhD support). It can also be possible to start your own group or to ask the institution under whose auspices you are doing research to employ an external, independent facilitator with whom you can meet, along with others experiencing the same sense of otherness. As Edwards puts it, make the invisible visible.

Reflective question

What have you learned about yourself as a researcher through your research practice? How has it helped shape your researcher identity?

Emotional labour

Emotional labour, a term coined by Hochschild (2003) to describe the performance that service industry workers practised in their interactions with customers, can equally be applied to researchers. Emotional labour recognizes that emotional detachment can be beneficial in effective interaction, so it is easy to see that for researchers feeling anxious about collecting data from participants, it can be useful (Frost, 2016). However, whilst this may be the case and effective for eliciting data, the hiding or denial of anxiety in other phases of research may not be as advantageous. Hiding anxiety because of feelings of being perceived as a fake, 'not doing it right' or being an inauthentic researcher can adversely influence the research but also adversely influence the researcher. Putting on the performance for fellow researchers, supervisors or team leaders risks you obscuring problems and challenges (in the research or with yourself). Conforming to the dominant perception of who a researcher is perpetuates it as the norm, minimizing your capacity to ever feel like you fit and conveying the same message to other researchers who may feel similarly. Being honest about how you are feeling, and why, with co-researchers can lead to useful discussions and can support others who may be feeling the same way to talk more openly about it. This could be of benefit to you and your co-researchers and definitely will be to the research because the infiltration of the anxiety is likely to be less as a result.

Reflexivity

Teaching yourself to be a reflexive researcher can be done using some of the strategies we have discussed above (self-reflection, making the invisible visible, writing). Adopting it as an approach throughout your career as a researcher will enhance your understanding of yourself, empower you to change and give voice (and support) to others. Journalling and writing field notes as your research progresses will open up insight into yourself as a researcher – the perceived barriers and opportunities – and will enable you to engage more fully with your research. It can be daunting both to recognize (perhaps unwanted) characteristics in yourself as a researcher that you may not have been aware of and to accept that there will likely be other characteristics that influence perception of you as a researcher by others. An openness to this can, however, be liberating. Whether you are a subjectively or objectively orientated researcher (or one that has to be both as in mixed methods research), recognizing that many aspects of being you influence the research and your enjoyment, or otherwise, of being a researcher. By working with (and sometimes silencing) your self-doubt, you are likely to minimize the chances of feeling disempowered as a researcher, freeing yourself to feel energized and excited by it. Rather than looking to others for what a researcher is, look to yourself in an honest manner that enables you to both learn from others and teach yourself.

'Researcher' identity

When being taught or trained as a researcher, we are often assigned a 'researcher identity' (Giampapa, 2011). It is formed from what we are told about ethical procedures, research processes and the expectations of roles and responsibilities of researchers. Quite necessarily, we are taught to prepare for data collection procedures, to learn different methods of data analysis and to maintain ethical standards that aim to minimize distress to participants. In practice though, researchers often find that they experience both personal and professional identities, either by claiming them or being positioned into them by participants. These can be identities which we feel comfortable with or which can provoke discomfort and unwanted emotions.

It may be that you choose to do research within a community that you belong to, assuming that your shared experience and understanding of the culture and nuances of the language will make it easier to access participants and the understanding of the research focus. You may choose to do research here for a variety of personal reasons: you want to give something back to the community, you want to help address these problems or you want to work with other community members in a mutually collaborative and educational stance (Yakushko, Badiee, Mallory & Wang, 2011). You may feel some guilt at the position of greater power and privilege that you have because of leaving the community for education and training elsewhere (ibid.). It can be upsetting then to find your motives for doing the research questioned by the community members or to feel that you are letting them down by not being able to fulfil requests for help that are not within your capabilities to give. Where you thought that being a researcher who is a member of the community would bring with it a recognition to your authentic desire to bring about change, you may instead feel rejected or unfairly questioned by community members. Whilst you may see yourself as an insider by race, class, ethnicity or any other shared characteristics with the community, as differences and perceived differences become apparent during the research, you may feel like you are seen as, and see yourself as, an outsider.

For these reasons, it can help to drop the dichotomy of whether you are an insider or an outsider researcher and, instead, consider yourself as *both* an insider and outsider researcher (see Chapter 9 for more on this). This can be on the dimensions outlined above and/or in intersections of your professional and personal identities.

Being armed only with your assigned researcher identity can leave you feeling misunderstood and rejected in your personal identity. Instead of being the researcher who wants to bring benefits through the research you are doing, you might experience being perceived with suspicion and mistrust – and feel it personally. In contrast, recognizing that the researcher you are can be intertwined with the researcher you present to others can be empowering and confidence-boosting. It can also help your research because you can open up dialogue and discussion with community members and participants, and mutual honesty can highlight the most appropriate research and how to do it. It will help to develop trust and rapport and foster collaboration. Recognizing what you bring, want to bring and are told to bring to research means that you can be a researcher who contributes, whilst recognizing and acknowledging that you are taking too from the participant group. As an insider and outsider researcher, you can claim your shared identity with participants, such as your backgrounds and upbringings, and the impact they have left on you with regards to your positions of power and privilege. You can also be open to yourself and the participants about differences – perhaps in dimensions such as gender, age, class and so on, but also, perhaps, in having lived different lives that mean your worldview has changed or differs in some aspects.

Self-reflection about who you are as a researcher can help you to recognize the personal identity that may mean you are bringing guilt or anger to the research. You can be better prepared for perceptions that question your agenda and how they differ to those of the participant group taking part in the research. It can help too to recognize the struggles and prejudices you have faced outside your role as a researcher because you identify as a member of a particular group. Sharing these and what it means to you to feel like a member (or not) of the group can help the participants to understand you better and offer insights into which aspects of their perceptions of you resonate with your understanding as well as which do not. By actively working with participants to identify perceived barriers to belonging that exist between you both, you can work towards building bridges across which dialogue can develop, helping you craft better research and create more acceptance of your researcher self. Being prepared to make and commit time to this will only be beneficial. Showing an authentic willingness to learn, help, collaborate and contribute through the research you are doing is likely to enhance this even further. Conversely, parachuting into the research field with unquestioned assumptions about the researcher you are is more likely to affect your sense of self when questions arise about your professional identity in your assigned researcher role. However, recognizing the identity you want to inhabit as a researcher can also mean you have to accept the contradictory perceptions of others and be willing to develop relationships based on this. The outcome is likely to be that you can better navigate your personal and professional researcher identities so that you then have agency over when to delineate them and how to recognize their intersection. This empowers you as a researcher, and participants will appreciate the associated openness. Such awareness can lead to better rapport so that in the flux of the research process, you are likely to be more confident as a researcher and do better research.

The following reflection demonstrates how professional and personal researcher identities can be provoked and, sometimes, become intertwined.

Researcher reflection (from an early career academic researcher)

During my initial readings of the data set, I was aware of my own positioning, and how this may influence my reaction to certain discourses. As a young person who also grew up during an economic recession, I tried to differentiate my own experience of the health service, education system and general adolescent development from the accounts of the participants. However, this subjective viewpoint did result in an emotional and empathetic response to the discourse, as individuals described their struggle to cope with various life stressors.

Power and privilege

A growing awareness of the influence of power and privilege on research has shown how marginalization can be reinforced, perpetuated or challenged. Whether the influence is on the research or the researcher identity, biased beliefs and assumptions about others can suppress and obscure voices and maintain prejudices. The power dynamics that exist between the researcher and the researched, between the research conduct and its dissemination and between the research aims and the audience it is intended for can impose unintended meanings and norms. Recognizing this encourages researchers to take a critical approach to understanding both their researcher identity and their research practice. Being proactive in self-reflection on their researcher identity, consulting and collaborating with participants and other stakeholders in the research and designing and justifying research designs and methods can highlight and reduce power imbalances. When trainers and teachers of research recognize their positions of power and privilege, and how inherent exclusionary approaches may be promoted to novice researchers, they encourage questioning and challenges that can help the shaping of researcher identities. New researchers are less likely to incorporate the beliefs and values that negatively impact their sense of researcher self and the research. Diversity in the research institutions and across researchers will increase the range of positionalities located about research practice and highlight issues of social injustice. There is a now growing awareness of barriers to inclusion and diversity in research practice, and researchers are increasingly required to consider their positionality in order to address them. This has come in part from those who recognize their positions of power and privilege and in part from researchers who are members of oppressed or marginalized groups.

Feminist researchers have done much to initiate and promote awareness of power and privilege by identifying how women's perspectives and understandings of their experiences have historically been obscured in research (Arnot, David & Weiner, 1996). Feminist research approaches highlight assumptions and taken-for-granted knowledge, in society, in research and in the connection between them.

One of the early feminist researchers was Carol Gilligan (1993) who highlighted gender differences in understanding moral development. She highlighted that moral development of girls and boys is different so that women have 'ethics of care' that influences their decision-making and behaviour, whilst men have 'ethics of justice', but this had not been recognized in the creation of dominant discourses about personal decision-making. New approaches to research subsequently emerged from questions about who the researcher is and what taken-for-granted assumptions are embedded in research. Women began to be included in research about them and their experiences and, over time, awareness of intersectionality grew as the multiple identities of women were highlighted. Instead of being categorized as one homogeneous group, women (and subsequently other groups) were seen as having differences across different dimensions (such as race, age and sexuality). The awareness has now extended beyond research with women, to research with other oppressed or marginalized groups, as was discussed in Chapter 9.

Awareness of power, privilege and positionality is not confined to subjectively orientated research. 'Strong objectivity' (Harding, 1995, 2013) is the counterpoint to apparently value-neutral research. It proposes that researcher bias always exists and that by placing emphasis on the researcher positionality, perspective and subjectivity (strong reflexivity), socially situated knowledge is generated. The researcher aims to conduct research from the perspective of those with whom they are doing the research, think meaningfully about their position about the research and challenge themselves in how it is taken up. Originally developed from concerns about sexism and androcentric bias in dominant scientific studies, strong objectivity requires the researcher to consider what lies behind the research questions and hypotheses so that bias, power and privilege can be recognized and addressed. Strong objectivity has now been incorporated into research ranging across business management and education, as well as in many research approaches.

Adopting strong objectivity enables you, as a researcher, to better recognize triggers and stages in the research when you may be exerting power and privilege or having it imposed upon you. Interactions with participants, what they provoke in you and how that influences the data collection and interpretation can be questioned, and unintended manipulation or imposition can be recognized. Interactions with supervisors and other more experienced researchers can be better understood so that you can recognize implicit bias and how it is influencing researcher identity and research practice.

It can be helpful too in recognizing that you may be affected by issues such as distressing topics or unexpected data and that the emotional impact and meaning can influence your positionality. For example, your perspective and positions in the research may fluctuate, and changes in your behaviour or the expression of emotions (some of which you may not be aware of) may be perceived by the participants. Their reactions and responses may differ in turn, as they detect shifts in power dynamics. Whilst you can prepare for the research and your role in it to a large extent, a lack of awareness and preparation for changes in affect which may impact your position can potentially lead to unscrupulous researcher behaviour (such as persevering with requests that are causing distress to participants, for example). Having some insight into your motivation and reasons for doing the research can help with this. If you are a researcher who is researching a topic because of personal experience of it (perhaps you have suffered with depression and wish to know the prevalence of people with depression in your community/age group/gender and so on, for example), you can prepare yourself to feel empathy with participants or share resonances with your own experience. Anticipate this and make plans for how you will respond, in the moment and after you have collected the data (perhaps with some mindful thinking or journalling, for example). Similarly, knowing that your positionality can change with the context and the interactions you have with participants, you may be more alert to differences in responses that may indicate differences in participant positionality. Being aware that, along with sharing commonalities with participants, you are likely to also have differences in views or behaviours can help to make sense of changes in responses or data that you find in the research.

Whether you are a subjectively orientated or an objectively orientated researcher, it can help to know that you will inhabit different positions during the research process. Choosing to make yourself as aware of these as you can (and recognizing that you cannot always expect the cues and positions to be conscious) will not only help you to recognize who you are as a researcher but also what has been assigned to you and what you have internalized. When dealing with challenges and barriers, this insight can be beneficial to both you and the participants and will make for research that is more rigorous, trustworthy and of value.

Chapter summary

This chapter has explored aspects of research identity and practice that are learned more from experience and self-reflection than from formal researcher training. They include feelings and emotions, such as excitement and anxiety, the role of relationships and support, and the awareness of positionality, power and privilege in the research. Ways of preparing for your experience of the research and ways of navigating it have been discussed so that your identity as a researcher can be better understood, your confidence as a researcher can grow and your research practice be more inclusive.

Further reading

McAlpine, L.. & Amundsen, C. (2017). *Identity-trajectories of early career researchers: Unpacking the post-PhD experience.* Palgrave Macmillan.

The book offers a synthesis of the empirically-based insights that arose from the experiences of 48 ECRs, who were participants in a ten-year qualitative longitudinal research programme. It examines decision-making processes underpinning the careers of PhD graduates and highlights the role of personal agency in negotiating academic and non-academic work and careers within broader personal lives.

Sarpong, J. (2020). *The power of privilege: How white people can challenge racism.* HQ.

This book is not written specifically for researchers but by offering practical steps and action-driven solutions based on understanding the roots of privilege and the systemic societal inequities that perpetuate it, it is useful to researchers who have been afforded privilege so that they can contribute to undoing the limiting beliefs held by society and help to build a fairer future for all.

Stewart, R. (2022). *How to do research: And how to be a researcher.* Oxford University Press.

This book provides a framework for developing and following a research career, or a career involving research, and provides a practical appraisal of the challenges and opportunities involved in being 'a researcher' in a wide range of academic fields.

11 Emotions and the Researcher

Introduction

Researchers are human, have relationships with other humans and gain understanding and knowledge about the world through cognitive understanding and the emotions they experience. When doing research, researchers can experience emotions ranging from anxiety, distress and frustration to joy and satisfaction. Acknowledging this is important not only to the well-being and support of researchers but to the direction and quality of the research. However, the attention paid to researcher emotions is much scarcer than what is paid to that of research participants, where the emphasis is on minimizing distress caused by taking part in the research. In this chapter, we will identify and discuss some of these emotions and what they can mean for you and your research. We will consider how you can support yourself and access support for managing them, and identify some ethical considerations associated with having feelings about and during your research. The chapter will end with a discussion on emotional challenges of being a researcher working alone and being a researcher working as part of a team.

What are emotions?

An emotion is 'a conscious mental reaction (such as anger or fear) subjectively experienced as a strong feeling usually directed toward a specific object and typically accompanied by physiological and behavioural changes in the body' (http://www.mw.com/home.htm). Emotions can be understood to differ from feelings because they are generated as sensations in the body, which are then interpreted as feelings. Think, for example, of being hot and uncomfortable in a social situation and then realizing that it is because you are angry about what someone is saying to you. Emotions can be short-lived or persistent, can come and go and can be provoked by internal or external events. Their location in the body means that there are often outward signs of them – shaking, facial expressions or sweating, for example – and, so, can be observed by others. Being conscious of your emotions does not mean you can adequately hide or suppress them, sometimes leaving you feeling embarrassed or vulnerable. As a researcher, you may find that emotions are directed towards the research process itself, participants or the data, regardless of whether you are conducting objective or subjective research.

The emotions that you have may be detectable by participants and can, perhaps, be transferred to them, therefore influencing the data. Similarly, of course, you may pick up emotions from participants too so that you both feel nervous or anxious. Emotional responses to data collection, analysis and interpretation are important too because they can influence how the process is carried out.

Anticipating and planning for emotional experiences as a researcher may seem obvious when you know that the topic you are researching is a sensitive or distressing one, but it is also the case that just about any topic is likely to evoke (positive or negative) emotions in you.

In the next sections, we will consider some of the emotional impacts of researcher vulnerability and collusion in research and, after that, consider how to prepare for and manage emotions.

Frustration, anger and the researcher

Frustration and anger can be provoked by the research itself, in interactions with participants and with yourself as the researcher. You can, at times, feel out of control over what is happening in the research process: there can be long waits for permissions and approvals to begin the research; there can be challenges in finding and recruiting participants; sufficient data may be hard to elicit and collect; meaningful results and findings may be elusive; and the writing up (and feedback on it) can make you feel inadequate and unskilled. When collecting data from participants, you can feel irate at what they have experienced, sad about the extent and effect of illness or bereavement and powerless to offer to help change in their lives. Sometimes, when participant emotions are powerfully expressed, such as if they cry or shout, you can feel the same emotions too, as well as guilt at having provoked them through the research. The rollercoaster of emotions you experience can lead you to being angry with yourself for not being a 'better', more caring or more sensitive researcher and induce a sense of vulnerability that can affect you and your research.

Personal and professional vulnerability

Vulnerability as a researcher is surprisingly common (e.g. Downey, Hamilton & Catterall, 2007; Emerald & Carpenter, 2015). Vulnerability is exposure to potential harm or attack. Ethical expectations require researchers to consider both the potential participants and the research topic in terms of whether distress is likely to be caused by taking part in the research. Vulnerable participants are those who may have a higher possibility of exposure to risk and/or increased susceptibility to harm and/or who may not have ability to safeguard their own interests, for example, due to impaired decision-making (Lajoie, Poleksic, Bracken-Roche, MacDonald, & Racine, 2020). Researchers are responsible for ensuring that participants have all the information they need to provide full and meaningful consent to take part in the research and must take steps to ensure distress is minimized throughout the research process (e.g. by offering opportunities to take a break or withdraw from the study). Participant vulnerability can be evoked by the research topic which requires them to provide information about themselves or experiences that are distressing or sensitive. The context in which the research is being carried out should also be considered. Perhaps, participants are within a community that experiences political, social or economic exclusion, have been ill-treated by research studies previously or have their data collected in environments that they are not comfortable in, which may include universities or healthcare settings.

It is true too, however, that researchers can be vulnerable for many of the same reasons. Listening to accounts of distress or observing concerning situations can cause a researcher to feel vulnerable about their own responses and their role. Researchers may feel they risk breaching cultural boundaries and causing offence if they are researching with communities that they are not a part of. They may feel an overwhelming sense of not giving back sufficiently or of the research not making things better for participants. Researchers may find it distressing to resist blurring their role as a researcher with a caregiving role or they may simply feel upset and angry about the experiences that participants are describing to them.

Vulnerability can manifest in different ways for researchers. Anger at the way participants have been treated or excluded can induce a sense of powerlessness. Being made aware of secrets that participants are withholding from others for their own protection can give researchers a sense of collusion with dominating oppressors. Sadness at experiences being recounted and documented by participants can become overwhelming, and sometimes lead to vicarious trauma for researchers (Coles, Astbury, Dartnall, & Limjerwala, 2014).

Researchers can also experience 'professional vulnerability'. This is vulnerability arising from the limitations and barriers placed on their research practice. For example, sometimes, participants and communities from which participants are drawn can inhibit researcher access to data collection. This may be because gatekeepers for the group do not support the research (perhaps, because of previous over-researching or bad experiences with researchers) and, therefore, restrict access to potential participants. Key actors necessary to the research may not engage with invitations to participate, and support for and publicizing of the research by elders and other gatekeepers may not be forthcoming. This can be distressing for researchers, who are keen to conduct the research, often on a timeline, but (obviously), this choice by the community representatives has to be respected. After considering whether it is because of you, the research topic or other aspects of the research process, it may be necessary to re-design your research, perhaps by re-locating to a different location or group. Finding out the community's history of being researched can help with understanding what may have happened before and why it has led to non-participation, as can ensuring you are aware of cultural practices and expectations to ensure you are not inadvertently breaching them.

Having overcome barriers to developing and beginning your research, it is also possible that participants choose not to engage fully with you in providing data. They may choose not to answer some questions or to hold back details. Whilst it can be easy to feel frustrated if this happens, it is worth reflecting on why this may be. There is a widespread trope, originating in feminist methodology, that giving voice to oppressed or marginalized participants is a way of enabling agency and empowerment. However, it is important to remember that those who choose not to speak out are doing so for their own safety or the safety of their community. Parpart (2010) highlights that in some societies, women who disclose that they have been raped are at risk of being killed by relatives, rejected by their community or divorced by spouses. Choosing not to speak out can be, therefore, more complex than a disempowering act: participant secrecy and silence can be a way of ensuring safety. Secrecy and silence can be distressing and frustrating for researchers who have dismissed it as a resistance or a choice not to be empowered, but understanding that the very survival of someone can depend on it can help contextualize the choice. Rather than seeing it as a missed opportunity for participants, the researcher can strive not to privilege voice over secrecy, and understand that steps to eventual empowerment may be small and many and the recognition of the role that secrecy and silence can play in this may be the best way to contribute to it.

Researchers seeking to access and interpret such silences should think their decision through carefully, with participant choice and well-being in mind. In addition, researchers may need to ensure safeguarding of participants before they take part in the research. Consent that is requested has to be detailed, sensitive and written in an appropriate and accessible language so that potential participants are fully aware of how anonymity and confidentiality will be addressed. When collecting data, researchers will need to gain the trust of participants and this may include reassuring them of their safety. Research goals need to be clear, and there should be allowances made for time, perhaps with repeated interviews, and for patience for detours and distress. The silences, whether verbal or expressed through body language, need to be carefully interpreted and, if possible, clarified with participants so that their distress is minimized. Of course, the research write-up has to be done in such a way that it continues to protect and anonymize participants so that they are not placed at risk of harm as a consequence of it.

It can also be that participants request that their data is withdrawn or that publication of the study is withheld. This may be for personal reasons or because it may unduly influence or prejudice other aspects of community life. Ballamingie and Johnson (2011) describe being requested not to publish their research because an important decision about a dispute in the community was being awaited. Participants who had campaigned for years were concerned that the study findings might influence the dispute outcome. Again, when such a request is made, the only ethical decision is to respect it. This professional vulnerability may be avoided by researchers clarifying participant views on research publication early in the process so that any potential barriers can be identified and addressed in a timely manner. For researchers needing or wanting to publish their research, there may be other aspects of the study that are ethically publishable – perhaps a paper on methodology or on other aspects of the study that are not related to the participants of that community.

Collusion

Collusion is 'a secret agreement or co-operation for an illegal or deceitful purpose' (Merriam-Webster). There is ongoing debate about deceit in research. Every step should be taken to avoid deceiving participants and it is generally only acceptable if no other non-deceptive method is appropriate, if the study will make a substantial contribution to knowledge, if significant harm or severe emotional distress to participants is not expected and if the deception is explained to participants as soon as the study protocol permits (American Psychological Association, 2016). However, collusion can leave researchers feeling guilty about their role in the research.

The rapport developed (and encouraged) in semi-structured interviews in qualitative research can lead to emotional responses by the researcher, and it has been shown that the time before and after interviews also provides researchers with feelings and emotional responses to participants (perhaps sensing their willingness to take part in the research and what they hope to gain by it) (Miller, 2015). This may particularly be the case when researchers are embedded within a community and using observation, interviews and focus groups to gather data. Living so closely, and sometimes for extended periods of time, with participants often means a strong rapport develops. It is not surprising then that when living with participants, rapport can include informal encounters during the fieldwork that can lead to friendships that endure beyond the life of the research project. Researchers describe wondering whether they have developed emotional collusion in these moments, in which they over-empathize with participants, sharing some of their own feelings and encouraging further engagement of their participants' emotions with their own. They question, afterwards, their motivation for this – was it to find relief and understanding of their own emotions or to elicit more data and insight into the experiences of the participants? Feelings of guilt about deceiving participants can then cause concerns about whether they should use insights gained this way in the research.

A large-scale research project using action research highlights how collusion can arise in other ways for researchers (Jones & Stanley, 2010). It is outlined below.

Research example (summarized from Jones & Stanley, 2010)

Study aims: to investigate the socio-emotional needs of pupils transitioning from primary to secondary school education in the UK and to develop, implement and evaluate pilot intervention programmes for three secondary schools and their feeder primary schools. It intended also to develop a training resource for teachers. The study aimed to be collaborative by using an action research approach in which activities were offered to primary school pupils that aimed to provide opportunities to get to know and interact with secondary school pupils six weeks prior to and after transfer.

Researcher team: The researcher team was made up of two university researchers (holding overall responsibility for the project), three teacher educators (to provide guidance and assistance to the classroom practitioners), senior classroom practitioners from three secondary partner schools (to manage the implementation of the activities in each school) and two local authority learning network co-ordinators (to act as a link between the secondary schools and the feeder primary schools). All team members agreed at the outset to contribute to the research strategies and evaluation tools and to play key roles in the collection and analysis of data.

Three key challenges: As the research progressed, however, it started to become clear that the stakeholders had different motivations, agendas and time constraints. The first challenge arose in finding ways to gain consent for taking part from the parents of the children involved. The Heads did not want to be involved in seeking consent for a variety of administrative, bureaucratic and role limitations. Hence, the researchers had to seek consent by contacting parents directly by post. Response rates were low, and the researchers were concerned that the diversity of cultural and linguistic differences within the pupil group was obscured by this approach. Nonetheless, they felt that they had to collude with the limitations of the heads' scope to exercise professional autonomy in order to ensure the project could continue.

A further challenge arose when one of the schools requested that a question be removed from the questionnaire before it could be distributed in their school, despite all team members being involved in the construction and approval of the questionnaire. Again, the researchers felt they had no choice but to comply, despite feeling that the filtering of information meant they were complicit in 'conformative' evaluation (Stronach & Morris, 1994, p. 5).

Writing up: When the research was written up and presented in a report to the funding body, the local authority and the three participating schools, the researchers subsequently learnt that disappointment had been expressed at its critical stance towards the evaluation and lack of sufficient celebration of the success of the interventions. The university where the researchers worked had an Initial Teacher Education partnership with the schools and authorities, so for fear of having a detrimental effect on this, they amended and reproduced the report with greater emphasis on the positive outcomes of the interventions. They felt that in colluding this way, their academic integrity had been undermined and this had subsequent influence on the ways in which they wrote papers for publication in academic journals.

The detail of this study shows several different ways in which these researchers felt that they had to collude during the research. Personal feelings of frustration, guilt and anger were provoked by the imposition of bureaucratic and structural expectations and demands. The researchers' obligations to carry out high-quality research whilst adhering to ethical and professional expectations of their role came into conflict with those of other institutions with which they were doing the research. Their compliance and collusion left them feeling like their beliefs and values as critical enquiry researchers were undermined and challenged, sometimes forcing them to amend the research and its write-up in ways that they were uncomfortable with. It demonstrates the importance of recognizing the possibility of having no choice but to collude when working with community stakeholders and what that can mean for how academic researchers feel about their research.

Some of what we have discussed in this section has referred to interviews and other extended face-to-face encounters with participants, but it is important to remember that all methods of data collection, analysis and writing up can lead to emotional distress and vulnerability. Even if exposure to participants may be more structured, and data collection not as focused on gathering detailed accounts as those of qualitative researchers, the effect of learning about the lives of others, however it is relayed, can lead to concerns on the researchers' part that the same could happen to them, influencing their own sense of risk and heightening their perception of threat from hostile world scenarios (Bluvstein, Ifrah, Lifshitz, Markovitz & Shmotkin, 2021). Whether objective or subjective, researchers' vulnerability can be increased because of conflict between participants' well-being and the limitations of what they can do to help as researchers. Whilst ethical requirements can enable support to be signposted to participants, there is little available for the researcher who may be left with an enduring sense of moral failure after their contact with participants has ended (ibid.).

There are, however, ways in which the vulnerability and emotional upset of researchers can be addressed, and we turn to these discussion in the next sections.

Reflective question

What are the emotions that you have experienced as a researcher? Are they more commonly positive or negative? What has provoked them in you?

Preparedness and resilience

Being prepared for emotional responses and reactions to the research you are conducting can help to manage and navigate them, even though it may not be possible to fully prepare for all the emotions you may experience or for what might trigger them. Remember, too, that the emotions may appear or carry through to after the research has been completed. However, it has been shown that researcher self-efficacy has been linked to confidence in carrying out research (Forrester, Kahn & Hesson-McInnis, 2004), so the better prepared you are, the better the quality of your research is likely to be.

Key components to being prepared are reflexive awareness of and resilience to your emotional reactions and having ways to manage them appropriately (note that 'appropriately' may mean acknowledging them to yourself rather than simply suppressing them). Preparedness has the additional benefit that it can help to identify what may be going on for the participants and, in turn, be useful in how you understand and interpret the data.

Self-reflection is an essential aspect of research. By thinking about your feelings and behaviour and what might lie behind them, you can better understand how you will engage and are engaging with the research process (reflexivity), what you are bringing to it and how you make decisions in it. In qualitative research, this practice is made explicit because of the recognition that your subjectivity will inform the research. Its benefits are not, however, limited to qualitative research because decisions about how data is elicited and analyzed are also personally informed in part. For example, decisions about what literature to read and tests to use will be influenced by your understanding of their contribution to the research, your knowledge of them and your confidence (and perhaps enjoyment) of using them.

Using self-reflection to understand more about your emotions can help you to prepare for aspects of the data that may evoke them. It may be the topic itself, perhaps chosen because of personal and professional experience or your preferences in collecting data – some researchers enjoy interviewing whilst others feel shy, for example. Knowing the motivations for your research and what you hope to achieve with it can also help you to better prepare for emotions you may experience during it. Perhaps, you feel strongly about a social injustice or want to research in a culture different to your own. Your emotions, feelings and your hopes for a beneficial change coming from it will be part of what is driving the research, You may anticipate feeling angry or powerless as it progresses and preparing for this in advance can help to mitigate and deal with these emotions.

Preparedness is also about reflecting on the power issues and inequalities that may exist between you and the participants and how they might influence your emotions. Ensuring that you have learnt as much as possible about the community or group from whom the participants will be recruited will help in approaching them with a degree of confidence that you are not going to embarrass yourself or be anxious about offending people. It also increases the likelihood that you will be given access to the group and that people will be more willing to take part in the research. Thus, you minimize experiences of feeling vulnerable and powerless as the research progresses.

However, no amount of preparation will negate the possibility of unwanted or unexpected emotions arising during your research. Being able to recognize them will help and finding ways to build resilience will enable better management and acceptance of them. Feelings such as concern and wanting to help can arise during the research of sensitive topics or research with marginalized or vulnerable groups. Associated emotions may be those of anger or shame that participants have had such experiences, and it can be tempting to want to give apology, advice or counselling in response. Knowing that this possibility may arise helps to build resilience to it by recognizing your boundaries as a researcher. Whilst having cognizance of your own emotions can help you to be empathic with participants by acknowledging that you are hearing the distress it has caused them, your responsibility is not to try and alleviate it for them but to ensure they are not caused further distress by your apparent indifference to it. For researchers who are working or training in support and care-giving professions such as nursing or counselling, the drift across the line from researcher to support-giver can be confusing and misleading to participants and realizing this can cause you to feel anxious or ashamed. Being clear with yourself about where your boundaries are as a researcher means that when you feel emotional responses that provoke a desire to offer support, you will be better able to have other strategies available, such as offering full debriefing, details of other resources participants can access and a follow-up welfare-check call.

Developing personal and organizational support can offer ways of building resilience. These may be in discussing possible reactions to participants or the topic that they are talking about with supervisors, mentors and team members. In turn, this can mean that emotions are less likely to take you by surprise if they do happen during the research and, if they do, that you have a support network to call on to discuss and debrief with.

The following reflection describes the powerful emotions that can be experienced from analyzing distressing data.

Researcher reflection (from an experienced academic researcher)

I found the sheer volume of data describing despair and confusion to have a strong impact on me. At times it felt close to overwhelming, and I had to take breaks away from it in order to return with a view to analysis, rather than succumbing to feelings of powerlessness. I was anxious that the young people writing the posts would struggle to find the help they wanted but that I could not help them to do so.

Emotional labour

Emotional labour is managing the outward presentation of your emotions when interacting with others in a job. The concept is derived from a study of people working in service industries (Hochschild, 2003) and may be seen to have a place when discussing emotionality in researchers. Whether it is that you feel overwhelmed by the despair of a participant or want to detach your-self from feelings of anger, the possibility that you can engage in 'deep acting' (in which you suppress emotions you are feeling) or 'surface acting' (in which you enact emotions and feelings you are not experiencing) can be an attractive one. It can protect you, encourage engagement with participants and enable more efficient data collection.

However, we can also consider that Hochschild suggested that emotion has a 'signal function' too, and this gives clues as to how to figure out other people's viewpoints and what is real for them. Seeking to hide or falsify your emotions during research can become a hindrance to the research process, both in the potential for inauthenticity that it may introduce and in partici-pants feeling that you are not understanding of or do not care about what they are telling you.

Objective researchers may find it useful to have emotion management strategies available to them, and subjective researchers may need to engage with them in order not to be overwhelmed during the research. Recognizing the emotional responses that you are experiencing and being prepared to share them with participants can help provide insight into the meaning of their experiences and give intuitive insight into their verbal communication (Hubbard, Backett-Milburn, & Kemmer, 2001). The reliance on, and inter-pretation of, verbal descriptions of experience in qualitative research can be enriched by considering not only the emotion expressed by participants but also your own emotional reaction to them. It may be that the knowledge expressed verbally is only rudimentary to understanding other people's experiences and by recognizing, supporting and responding to the emo-tionality, you obtain a deeper and richer understanding. This requires an openness to your own emotionality so that you can take agency over how much of it is appropriate to share, when it is valuable to suppress it and how much of it is a mirroring of the participant. In all instances, it is useful to reflect afterwards on your emotional experience of the research, perhaps taking time alone to allow yourself to fully respond to feelings that have been aroused, whether they be anger, sadness or even happiness at how the data collection went.

Use of emotional labour strategies may also be influenced by your personal characteristics. If you are a shy researcher, nervous of recruiting and collect-ing data from participants, it can be helpful to acknowledge this in advance and learn to suppress your anxiety when you are engaging with participants. If you are a researcher working to tight deadlines or have encountered setbacks and challenges in the research to date, you may be nervous about collecting sufficient and appropriate data in time. If this is communicated to the partici-pants, they may pick up on it and feel inhibited about giving you data. Or it may be that in the course of the research, you hear about experiences that are close to your own (recent bereavement or diagnosis of serious illness, for example) and you may be susceptible to imposing what you are feeling onto what the participants are describing to you.

It is a key priority of all research to respect participants and acknowledge the privilege of them being prepared to share information about themselves with you. In terms of affect, one of the best ways you can ensure respect and gratitude is to prepare well for whatever emotions you might encounter, in the participants and in yourself, understand why they might arise and recognize them when they do. This will not only help the research process but also ensure that you can reflect and debrief afterwards with yourself, your peers, and your mentors and supervisors. Not doing so risks tainting the research with unknown, and unrecognized, elements that, in turn, can lead to thin data and misinterpretation of it. On the other hand, doing so will enhance the researcher–participant relations, raise the quality of the research and your confidence as a researcher, and probably mean that you enjoy doing the research more.

Researcher emotionality in teams

The privacy and unique experiences of emotions can mean that researchers are reluctant to share them with co-researchers. There may be a sense of failure or a vulnerability to being criticized for reacting in unexpected ways. However, it is useful to remember that all members of the team will have emotions and that they may experience them too during the research. Being open to sharing and seeking support from co-researchers with whom you feel safe can be empowering for you and an invitation to them to do the same. Research teams can be diverse – we have seen, in the previous research example, a description of a research team made up of stakeholders from different public community settings (Jones & Stanley, 2010) and teams of researchers can also be drawn from the same university, laboratory or clinical institution. The range of research experience of the team members is likely to vary, and it may be the case that it feels more appropriate to approach some members than others (perhaps, those with whom you work on other projects, or teach with, rather than members who are new to research or not known to you, for example).

Research has shown that emotionality can be experienced in several roles of the research team members (Sikic Micanovic, Stelko & Sakic, 2019), that is, not only those who collect data but also those who transcribe, code and analyze it. Repeated exposure to stories of distress and sorrow or ongoing collection of data about the frequency of sensitive or distressing events are likely to have an emotional effect, even if face-to-face contact is not part of the data collection (e.g. Sherry, 2013; Warr, 2004; Woodby, Williams, Wittich & Burgio, 2011). The cohesion and performance of the team are likely to be increased if there is an openness and authenticity between team members that allows for safe sharing of emotional experiences and the impact of doing research.

Emotional intelligence has been shown to positively affect the success of teams, and this can be considered as applicable to research teams too.

Emotional intelligence

Drawing on the concept of the existence of 'multiple intelligences' (Gardner, 1983), emotional intelligence is monitoring and regulating your own feelings and using them as a guide to your thinking and behaviour (Mayer & Salovey, 2007). The concept has interpersonal and intrapersonal intelligences as its basis (Luca & Tarricone, 2001) and identifies five main domains: knowing one's emotions, managing one's emotions, motivating oneself, recognizing emotions in others and managing relationships. A further adaptation (Goleman, 1998) developed five emotional competencies that make up emotional intelligence: self-awareness, self-regulation, motivation, empathy and social skills. It is useful to consider each of these in the context of being a researcher who is part of a team.

Self-awareness is the ability to understand and interpret one's own feelings through internal reflection (Luca & Tarricone, 2001). Internal reflection is valuable, and often essential, to researchers because it enables critical insights into how you interact with others and gives you the opportunity to change your behaviour towards them. This can help with developing rapport with participants and productive relationships with fellow researchers.

Self-regulation is knowing and understanding your emotions to enable opportunities to regulate them. This is useful to researchers because you can focus better on the relationships you are building with participants and fellow researchers by navigating tensions or managing unexpected or unwanted emotions.

Motivation helps in being focused and proactive with perseverance and initiative. This helps researchers as they experience both positive and negative emotions and can support fellow researchers to feel motivated too.

Empathy is the capacity to recognize other people's feelings. To recognize feelings in other people, you must first recognize these feelings in yourself. This is helpful to researchers because it can demonstrate understanding and support to participants and fellow researchers.

Social skills are necessary for effective communication and effective relationship development. For researchers, they are important in interactions with participants and can help to navigate conflict and tension with fellow researchers in the cultivation of productive and supportive relationships.

Emotional intelligence is a valuable tool for researchers and requires a high degree of self-reflection. It means developing reflexive engagement with the research process and the people you encounter during it. Emotional intelligence can be enhanced through writing to critically recognize and question your emotions and the behaviours that they induce, having meaningful communication with others and being open to receiving feedback on how your behaviours are interpreted. Practising reflexivity in research teams can bring valuable insights into both your interactions with team members and the research itself. It allows questions about your research practice to be asked and for understanding of what other researchers are bringing to it. However, it can be challenging and evoke vulnerability and anxiety.

As a researcher who is part of a team, you are bringing your expertise, knowledge and experience to it. This may be from your training and education, from the community you are part of or as a stakeholder with interests in the topic besides academic. The composition of multiple perspectives enhances the creativity and depth of the research but can also lead to confusion and concern about your own contribution or that of other researchers. It may be that you have been invited to the team because of your methodological expertise to enhance and support other methods being used in the group. You may not have been previously aware of other methods used, as fellow team members may not be aware of yours. A reflexive team approach will enable questions to be asked that can clarify and explain the value of each method. In having questions asked of you by other team members, you have the opportunity to consider, again, the rationale for the choice of your method, how you are using it and – when it comes to the analysis and interpretation of the data – whether assumptions and biases have been acknowledged and addressed. Contradictions, gaps and unintended impositions of meanings may be identified and explored, rather than obscured or cast aside. Open communication about method choice and use can also help to alleviate misunderstandings or frustrations about different timespans in carrying out the analysis or reporting the results.

As a team member looking to interact productively and supportively with others in a reflexive manner, you may also need to be prepared to discuss your motivation for being part of the project team. It may be that this can sometimes feel exposing and personal (Frost, 2016) or it may raise concerns amongst the rest of the team. Being clear with yourself about why you are taking part, and acknowledging to yourself what you wish to gain from it, can be helpful when providing explanations to others. If, for example, one of your aims is to explore an experience that you have had in your personal or professional capacity, it is useful to reflect on why you think it valuable to explore similar experiences of others. Knowing that there may be emotional trigger points in the research is useful when it comes to discussing it with co-researchers, who, after all, may have similar motivations and may find your openness useful in acknowledging this to themselves and others in the team.

Team-working and reflexive practice may be challenged if the researchers are in different geographical locations. Regular, face-to-face meetings may not be possible and you may feel isolated with fewer opportunities to address challenges and develop new ideas. This can be addressed with technology – both in the arrangement of 'formal' team meetings, and if appropriate, with additional support available through contact with team leaders and other researchers (perhaps, ones who are carrying out the same role as you in their location).

Part of all research is its writing up and dissemination. If you are a community researcher, it may be more important to you that the research is made available to the community, whilst academic and practitioner researchers may need to have research published in academic journals. Early discussions about who will lead the writing up, what the contributions of each researcher are expected to be and how the research will be disseminated are likely to reduce upset and confrontation later in the process.

Emotional intelligence used in a research team context will not only enable you to be more aware of your own motivation, but is also likely to raise your confidence in your ability to interact with other team members. Through clear communication that respects the contribution of others, you will be able to clarify what is expected of you and of them. Adopting an open communication style that shows a desire to learn and empathize with the professional and personal aspects of conducting research together is likely to prepare you better for challenges on emotional and practical levels and to ask for and give support to team members who may encounter the same. If you feel that your contribution is being marginalized or overlooked, you will be better able to raise this before resentment sets in and discuss possibilities for why this has happened (or you think it may happen). With a commitment to self-reflection, it is likely that possibilities of conflict or disgruntlement can be reduced and addressed, which, in turn, will mean that the research itself will be of higher quality. Above all, you may form meaningful relationships with co-researchers from whom you can learn (and maybe collaborate with again in the future), your passion for the research will be more evident and it is likely to be a more enjoyable experience for everyone involved.

Chapter summary

In this chapter, we have considered what it means for researchers to experience emotions when carrying out research. We have discussed that emotions can arise in anticipation of the research as well as during it and can also arise when the research is completed. We have identified that emotions can be directed towards the research process itself, in interactions with participants and towards yourself. Being prepared for the likelihood of experiencing emotions can be helpful in developing resilience and ways of managing them, and the chapter has considered the roles of emotional labour and emotional intelligence in helping with these processes. Recognizing that emotions may arise for you as a researcher will not only help your well-being but also raise the quality of the research, whether you are an objective or subjective researcher.

Further reading

Flam, H., & Kleres, J. (Eds.). (2015). *Methods of exploring emotions*. Routledge.

This book of short essays contributed by researchers from a range of disciplines provides a range of perspectives on studying emotions. The first person accounts include reflections on the effects of researcher feelings and touches on the ethics of emotions research.

Loughran, T. & Mannay, D. (2018). *Emotion and the researcher: Sites, subjectivities, and relationships*. Emerald Publishing Limited.

This book is a collection of contributions from a range of well-established and ECRs, who engage with the emotional experiences of researchers working in different traditions, contexts and sites and demonstrate their centrality in data production, analysis, dissemination and ethical practice.

Ryan-Flood, R., & Gill, R. (Eds). (2010). *Secrecy and silence in the research process: Feminist reflections*. Routledge.

This book is a collection of essays by international researchers in a range of disciplines. It considers debates and questions about epistemology, subjectivity and identity in research and provides first-hand insights into how researchers have navigated difficult dilemmas about who to represent and how, what to omit and what to include.

12 Researching Researchers

Introduction

Being researched as a researcher means your research is monitored and evaluated and your researcher characteristics, such as collegiality, ability to communicate the research and reputation with other researchers, are considered. It is done to assess your contribution to applications for funding, promotion and further employment. The research may be by institutions, or individuals, and the information gathered is key to important decision-making in researchers' careers. This means that not only is meeting research practice expectations significant but so is gaining the respect of others. In this chapter, we will consider how you can enhance your profile as a researcher, both in terms of your research and of yourself as a researcher. The chapter starts by giving different definitions of long-term and career researchers, before moving on to describe formal assessment structures. It then discusses expectations placed on researchers, such as gaining funding and research publication, before considering the role and importance of personal characteristics. The chapter ends with ideas and tips for effective dissemination of your research.

Career researchers

Career researchers are those who develop experience, expertise and promotion to enable being a researcher for a significant proportion of their career or as their full-time profession. Chapters 4, 5 and 6 have discussed some of the pathways to this. The trajectories take many years and you will spend differing numbers of years in different positions, sometimes plateauing and sometimes moving through more quickly. Several countries have put time limits on post-doctoral research positions to try and progress careers. For example, the EC Marie Sklodowska Fellowship defines postdoctoral researchers as being within eight years of a PhD being awarded, and Germany states that postdoctoral researchers can be on contract for a maximum of six years, after which they must be offered a permanent contract or made redundant. However, recent research (Menard & Shinton, 2022) has found that despite this, many postdoctoral researchers are in the role for longer than the stated times. In the UK, for example, 34 per cent of UK researchers remain on temporary contracts for over 10 years. In addition, the majority of universities in the UK and Europe do not have research-only pathways which can mean job insecurity, lack of status in the academic career structure and poor working conditions. The situation is similar across Europe.

Moves are being made to address this, including the European Commission proposal for a new framework for research career structures for its member states (ec.europa.eu, 2023). The set of measures includes a Council Recommendation to establish a new European Framework for European Careers, a Charter for Researchers to replace the 2005 Charter and Code for Researchers and a European Competence Framework for Researchers to support inter-sectoral researcher mobility. The initiatives seek to develop accepted definitions for researchers and research professions and to highlight the importance of research management roles. The Framework recommends an increase in permanent contracts for researchers and for working conditions to be improved. The proposal has been made in light of increased recognition of challenges of job insecurity for researchers, particularly those at the start of their careers. Its objectives are both to retain researchers in the EU and to make it an attractive place for researchers to work. It includes an investment strategy to fund a consortia of funding bodies with jobs for early career researchers, in order to make research careers more attractive and to create a mini-job market for researchers. The proposal and investment strategy have been put to member states and welcomed in a joint statement from the Aurora European University Alliance, the Coimbra Group of multidisciplinary universities, the European University Association, the Guild of European Research-Intensive Universities and the Young European Research Universities Network. These bodies caution that this is only a first step and actioning it will be voluntary.

So, although there is an increasing focus on improving conditions and enabling career progression for researchers, being a researcher is not a straightforward career path to follow. Menard and Shinton describe career researchers as long-term researchers (LTRSs) and conducted a case study of LTRSs in a UK university to highlight some of the challenges and categories that researchers find themselves in. See Table 2 for a summary of these.

This picture of those who want to be researchers was formed from interviews with LTRS in one UK university, but we can expect similar challenges and levels of status to be replicated in other countries. When set against statistics that show that 38 per cent of researchers in the UK covet research-only positions (Mellors-Bourne & Metcalfe, 2017, cited in Menard & Shinton, 2022), it seems that becoming a researcher after your PhD is not a short-term transitional stage or a stepping stone but a role in which you can remain for many years. However, with the increasing awareness of the challenges to researchers choosing research as a career, their own determination to pursue careers doing what they love and the recognition of the value of LTRS, perhaps it will slowly change.

Table 2 Types of long-term researchers (adapted from Menard & Shinton, 2022)

Researcher category	Description	Challenges
'Candidate' LTRS	These researchers are planning on becoming academics and so apply for posts in the chronological order, becoming researchers/research fellows after their PhD with a view to further applications for lectureships.	Usually on fixed-term contracts Cannot be principal investigators (PI) for most funders (because of fixed term contracts) Usually have to organize own teaching and management experience, necessary for lectureship applications May have added pressure to demonstrate more publications and in higher impact factor (IF) journals Can be demoralizing to spend years 'plateauing' rather than advancing
'Accidental' LTRS	These researchers have not planned to be long-term researchers but become LTRS for different reasons: Work for research groups(s) that are successful in obtaining funding to enable multiple research projects, sometimes with the same PI over years May have sought out the university, the group or the PI because of their good reputation Accidental LTRS without PhDs	Not eligible for fellowships or lectureships Often motivated by financial income need rather than expertise

Table 2 Types of long-term researchers (adapted from Menard & Shinton, 2022) (*Continued*)

187

	Interdisciplinary accidental LTRS, who have extensive knowledge across disciplines	Unable to obtain discipline-specific fellowships or lectureships because regarded as not specialized enough
		Job application reviewers are usually discipline-specific, so not able to holistically review the application
		Often regarded as 'not good enough' to be full-time researchers
	Part-time accidental LTRS: in this study, all were women choosing to work part-time, to fit around other commitments	Expected to carry out full-time duties in part-time hours
		Excluded from some tasks in belief it is helpful
		Less time to produce publications
Career LTRS	Those who actively choose to become LTRS, some leaving academic posts to do so and others after being 'accidental' researchers	Not a choice that can be made at the outset of one's career because of lack of research-only pathways
	Have extensive and wide-ranging knowledge from working in different institutions	Not on permanent contracts, so cannot be named as PIs on funded research
	Adaptable across projects and disciplines	
	Often have good reputation and extensive networks	

Reflective question

Where does your role as a researcher fit into the definitions in Table 2? Is this by design or evolution?

Assessment of universities

Most universities are funded by taxpayers with decisions about the money that they should receive from government commonly made by Performance-based Research Funding Systems (PRFSs). The purpose of these is to assess and evaluate research so that funding can be allocated accordingly.

An example of a PRFS system is the UK-initiated Research Excellence Framework (REF) (ref.ac.uk). This scrutinizes how universities show accountability for public investment in research, produce evidence of the benefits of this investment, provide benchmarking information and establish reputational yardsticks. In turn, this information is used within the higher education sector and for public information, and informs decisions about allocation of funding for research.

The assessment is undertaken every six years by the four UK higher education bodies: Research England, the Scottish Funding Council, the Higher Education Funding Council for Wales and the Department for the Economy, Northern Ireland. The REF aims to:

- provide accountability for public investment in research and produce evidence of the benefits of this investment
- provide benchmarking information and establish reputational yardsticks for use within the higher education sector and for public information
- inform the selective allocation of funding for research.

The process of expert review is carried out by panels made up of senior academics, international members and research users for 34 subject-based units of assessment. Assessment is made of the quality of outputs, such as publications, performances and exhibitions, their impact outside academia and the environment supporting the research. It combines performance-based institutional funding with research evaluation and, in doing so, is unique amongst other systems in Europe. The inclusion of review by peer researchers enables both qualitative assessment and quantitative evaluation. This approach contrasts with some European countries, such as Denmark, Finland and Norway, which use indicators of institutional-performance only, rather than panel evaluation *and* peer review, to make funding decisions. The few countries that use both panel evaluation and peer review either do not use them for funding decisions or replace them, in part, with performance-based assessments. Sweden and the Netherlands organize assessment exercises within (and to some extent by) each university, and these do not include funding implications. They are co-ordinated at national level by a Standard Evaluation Protocol (SEP). The PRFS system is autonomous and self-evaluating, and performance indicators representing research are not part of it. Norway's and Portugal's assessments have more of a formative and advisory function than the REF does. The increased flexibility brought about by breaking the link with funding allows universities in these countries to be more flexible and to evaluate themselves with a more thematic rather than institutional focus.

Importantly, whether the assessments combine research evaluation with funding allocation as in the UK, or carry them out separately, researchers and their research are themselves researched for their performance and output. This places publishing your research high on the list of aspects of your work that will be scrutinized. In the next section, we will explore what this can mean.

Publishing in academic journals

It is important that research reaches both those who it is hoped will benefit from it as well as other researchers, so that it helps to advance the field. Publishing research enables this. Non-academic audiences are more likely to read research reports and presentations than academic journals, but disseminating research in journals remains an important strategy. Academic journals are published periodically and aim to make research available to the scholarly community. Researchers submit their research write-ups as manuscripts, usually ranging in length from between 4,000 to 8,000 words. The manuscripts are blind reviewed by peers in the field who assess their content, value and interest to the journal audience. The aims and scope of each journal are included on their webpage and help researchers assess the prospective interest in their choice of research topic and approach. The journal editor will determine whether the manuscript is likely to be suitable for the journal and whether to send it out for review. A 'desk-rejection' by the editor means you can then submit to another (more appropriate) journal. If sent for review, however, authors are then invited to use reviewers' comments to strengthen the manuscript and resubmit the manuscript for further consideration for publication. A manuscript can only be submitted to one journal at a time, so the timescale from writing it to seeing it published can be lengthy, particularly if you submit to successive journals. It is, therefore, important to inform yourself thoroughly about the focus and style of articles that each journal publishes before submitting. Once accepted, the manuscript is sent to production editors who will return proofs to you for approval before publication.

As rewarding and important as publishing your research is, the process can be challenging as well as time-consuming. Receiving what can feel like negative feedback on your carefully crafted, and recrafted, manuscript can feel hurtful or frustrating. Authors sometimes feel that the point has been missed by reviewers or that their comments are unduly harsh. It can be tempting to give up on getting the manuscript to a publishable standard. Good advice is to leave some time from receiving feedback to addressing it if you feel this way. Sometimes, it is worth considering whether to find another journal, although this should be a carefully thought-through decision because there is no guarantee that there will not be amendments required from a new journal. Above all, remember the comments are not directed to you as a researcher but to the written article – the feedback may include highlights of important omissions and requests for clarification of details or developing arguments, and they usually help to make a stronger, publishable paper.

Journal impact factors

The impact factor of a journal is a numerical measure of the number of times an article has been cited in the last two years divided by the number of articles published by the journal in the last two years. The impact factor index is calculated by Clarivate, an analytics company. Impact factors are useful to researchers because they help to decide where to submit a manuscript for publication to make the most impact. They are also used by universities to help to make decisions about researcher promotion or tenure. The journals with the highest impact factors tend to be medical and scientific journals; *Cancer Journal for Clinicians* and *Nature Reviews Molecular Cell Biology* were ranked numbers 1 and 2 in 2022 with impact factors of 508.702 and 94.444, respectively. However, it is important to remember that publishing your research in a journal widely read by researchers with similar (perhaps niche or less mainstream) research interests to yours is as valuable in disseminating your research.

h-Index

The h-index is the number of publications a researcher has authored that have been cited at least h number of times.

i10-Index

The i10-index is the number of publications a researcher has authored that have been cited by others at least 10 times.

Open access publishing

The movement to make scholarly publications freely available began in the 1990s with the initiation of open access publishing. It followed the previous Open Science Framework (osf.io). It is based on the premise that publicly funded research should be available at no cost to all those who want to read it. There are no financial, legal or technical barriers to anyone wanting to read, download, print, distribute or search for information. For researchers, it has the advantage of increasing the visibility of their research and of the likelihood of the reuse of their research results and data.

In 2003, in response to the rapid availability of the worldwide web, the Max Planck Society published the Berlin Declaration on Open Access to Knowledge in the Sciences and Humanities. It defined open access contributions as '... original scientific research results, raw data and metadata, source materials, digital representations of pictorial and graphical materials and scholarly multimedia material' (https://openaccess.mpg.de/Berlin-Declaration).

The Declaration defined open access contributions as meeting two conditions:

> The author(s) and right holder(s) of such contributions grant(s) to all users a free, irrevocable, worldwide, right of access to, and a license to copy, use, distribute, transmit and display the work publicly, and to make and distribute derivative works, in any digital medium for any responsible purpose, subject to proper attribution of authorship (community standards will continue to provide the mechanism for enforcement of proper attribution and responsible use of the published work, as they do now), as well as the right to make small numbers of printed copies for their personal use. (ibid.)

Written permission to do this has to be provided. A complete version of the work and all supplemental materials (and the written permission) in an appropriate standard electronic format is deposited (and thus published) in at least one online repository using suitable technical standards. The repositories must be 'supported and maintained by an academic institution, scholarly society, government agency, or other well-established organization that seeks to enable open access, unrestricted distribution, inter-operability, and long-term archiving' (ibid.).

Most universities now have a repository in which you are expected to deposit your PhD thesis and subsequent research.

Open access publishing has many advantages to researchers – increased visibility and having your research read by a more diverse audience – but there can be disadvantages too. Although freely available to audiences of your research, the costs of publication sometimes have to be borne by the researcher/author themselves. Known as the Author Processing Charge (APC), publishers transfer the costs from the audience to the authors. APCs range in cost, depending on the journal and the articles it publishes. The average cost for an article is €2,000. Preparing graphs and other figures can increase the cost. The payment is due once the article is accepted by the journal and before publication. It is important to check the journal website prior to submitting so that you have an idea of whether it will charge APC and how much it is likely to cost you. It is also possible to make a case to the journal on an individual basis, arguing perhaps that your institution does not have the funding, the paper has particularly widespread interest in your country or that your university is in a developing country. Some research funders (e.g. the Gates Foundation) include costs for OA publishing and some universities will pay, but this is not always the case, so it is important to check with your institution before going down this route.

Creative Commons (CC) Licences are a standardized way to transfer copyright permissions for your academic research. It is necessary to give this permission for your work to be published as open access. There are different categories of the license. These are outlined in Table 3.

Table 3 Creative Commons Licences (adapted from https://creativecommons. org/share-your-work/cclicenses/)

Type of license	Permission type
CC BY	Permission for the public to distribute, remix, adapt and build upon the work as long as credit is given to its creator.
CC BY-SA	As above and allows for commercial use. Any adaptations from reuse must be licensed under the same terms.
CC BY-NC	As above but only for non-commercial use.
CC BY-NC-SA	As above plus any adaptations must be re-licenced under the same terms.
CC ND	Material can only be distributed or copied in unadapted form. Allows for commercial use and credit must be given to the creator.
CC BY-NC-ND	Material can only be copied and distributed in unadapted form and only for non-commercial purposes. Credit must be given to the creator.

The CC BY licence is the most liberal and the one closest for giving permission for your work to be made open access.

Plan S was developed by a group of national research funders, European and international organizations and charitable foundations with the aim of making full and immediate open access a reality. cOAlition S was developed around this to focus on research publications. Members are committed to making all peer-reviewed publications arising from research funded by them openly and widely available. It states that from January 2021, all research funded by them must be published as open access. The coalition will financially support the publication of work in open access journals or on open access platforms. They will not support hybrid open access fees in subscription journals but state that publications should be made immediately available in open access repositories.

Some journals, such as the *European Journal for Qualitative Research in Psychotherapy*, are fully open access (the 'golden' route to open access publishing), with publishing costs met by sponsoring organizations. The Directory of Open Access Journals (DOAJ) lists fully open access journals across 136 countries and in 80 languages. Other journals are hybrid-subscription journals with the option of authors paying APC to have their article published as open access (the 'green' route to open access publishing).

With the rapid rise in demand for fully open access journals, many previously subscription-only journals have transitioned to becoming hybrid or fully open access. As a researcher looking to have your research published, however, it is worth considering the ranking of the journal alongside its APC costs.

Journal ranking works in favour of journals that have been published for longer because impact factors are calculated over the previous two years (see impact factor definition above). Fully open access journals tend to be newer and, so, will have lower rankings until they are established. To address this, these journals sometimes increase the number of articles included in each issue, and some academics have questioned the review process and its impact on quality of articles that this has led to. However, it is also the case that some top-ranked journals, such as the suite of Plos journals, are fully open access.

Moving from subscription to hybrid open access, on the other hand, means that a journal keeps its ranking. It may be important to you to publish in a high-ranking journal, but it may also be that the cost of paying for your article to be published as open access may be greater than publishing in a fully open access journal. Look out too for predatory journals. These journals have reacted to the new publishing model predicated on open access publishing by charging authors but not vetting the quality of the papers submitted. Publication (at a cost) is guaranteed but with the lack of or minimal peer review, there can be great variation in standards. This is not only damaging to the reputation of researchers but also means that readers veer away from consulting these journals; your research may be very good but paying for it to be widely available may backfire if the audience is limited. Thinkchecksubmit.org enables you to research trusted journal and book publishers before deciding where to submit your work.

Writing for journal publication means you must convey the key point(s) of your research in a succinct style with appropriate detail. Readers must be able to understand how the research was carried out so that they are able to replicate it, and it must be clear how the results were reached. This is a different approach to thesis writing, so researchers new to writing for publications will benefit from seeking feedback from supervisors, mentors and other experienced writers. Learning to become a reviewer for journals helps you to familiarize yourself with the writing style and format of different journals. It's also useful to check journal information for guidance on what is expected of the writing, and some journals include occasional editorials detailing how to write research carried out with particular methodologies.

The following reflection from Finlay (2020) describes the relationality between embodiment and writing.

Researcher reflection (from Finlay, 2020)

When I sit down to write, I often begin with doing some actual Focusing (Gendlin, 1996). I tap into my body wisdom and 'feel' my topic from the inside. Sometimes, something important or interesting emerges out of those moments of embodied reflection. I might experience excitement bubbling up about a particular line of argument – excitement that tells me that 'this' is what I need to write about. I try to follow my motivations and interests. I open myself to what pressure I'm putting on myself and perhaps recognise some of the expectations from others. As I begin to see what it is that I need to do, I become aware of some of the traps ahead

Books

When you are being researched as a researcher, your university or research institute may consider whether you have written or edited books. Some disciplines expect their researchers to write books whilst others expect more journal articles. In the middle are disciplines such as law or psychology where publishing books is not essential but can help with assessment of you as a researcher. The decision to write a book is a complex one. A book is more likely to reach a wider and bigger audience than a journal article will, and as a book author, you will be freer to consider material from across disciplines or outside your research approach. You may be motivated to write about your research in ways that cannot be described fully in a short journal article because you want to develop and expand arguments in a more detailed way than the word count of an article will allow. You can also be more flexible with the format than when writing a journal article, although it can be challenging too to write in a sustained style over the length of a book. The time taken to write a book means that you may not be able to continue writing journal articles during the book-writing period or that the book takes many years to complete because of other commitments. Writing a book can be enjoyable, however, offering a chance to reflect on your research and learning about other research that you may not have otherwise come across. The writing process itself can be stimulating and bring new insights to your work.

There are other issues to consider when contemplating writing a book. You will need to develop a proposal for it for the publisher and then reconsider it in light of the comments from reviewers that the publishers will send it out to. Once you get going, there may be little contact with your publisher but deadlines will have been set and agreements made about draft chapters. It is good to stay in touch with your commissioning editor to keep them updated and to warn them of possible delays in meeting deadlines. Although the financial income is likely to be small, it is important to inquire into royalties that different publishers offer and to check your contract for issues such as translation and foreign language rights. You want to be clear too whether the book will be published as hardback, softback and/or e-book. For first-time authors, the royalty percentage of the sales of the books is usually in the 4–5 per cent bracket. If you go on to write more books and to become recognized as an author in your field, the royalty percentage may increase, to a peak of around 10–12 per cent. Money is, therefore, not often a primary motivation for academic book writers!

Once the book is finished, there is a lengthy process taking it to publication. This includes having your manuscript reviewed, and editing and amending the book in response, compiling (or having compiled for you) an index, proofreading the manuscript, choosing a cover, seeking out other researchers who may be prepared to endorse the book and writing any acknowledgements or dedications not yet included. This takes a few months of communication with your publisher until finally the book is published. Factoring in the time to write a book is, therefore, a decision that should be taken with regards to how it will be evaluated by your university department, what other commitments you have and your personal life requirements. For those who do write academic books, however, the process is rewarding and can help establish your reputation as a researcher.

Sole authoring/co-authoring

Whether writing for journals or writing a book, the question of whether to do it as a solo author or with others is an important consideration. One of the qualities considered important in research is that of collegiality and good communication. This can be evidenced in examples of co-authoring or editing a book. Writing with others helps to bring different insights and expertise and enables the group of writers to review and critique as the writing progresses. It does mean, however, working to timetables that suit you all, so you may have to go faster or slower than you would if you were writing alone. Depending on how you have structured the group, decisions may have to be made democratically and consideration given to inclusion of material you might not have considered if writing alone.

Conversely, writing alone gives you more agency over how you plan and write the book, and setting personal and publisher deadlines can be easier. It can, of course, take longer and the responsibility for its completion lies only with you, which can be challenging if you suffer from writer's block or simply don't have as much time to give it as you would like. In making the decision about authorship, you need to think about not only what is best for the book but also what is best for you and what status your discipline places on sole vs co-authorship.

Having discussed how your writing as a researcher will be assessed, the next section turns to another common focus of researcher evaluation – attracting research funding.

Gaining funding

Applying for funding for research is a time-consuming process and, of course, does not always result in success. However, most evaluators of your promotion prospects will be interested in how you evidence your ability to do this.

As you start on your career as a researcher, it can be hard to evidence that you and your research are worthy of being funded. Establishing a track record both of publications and previous funding helps to demonstrate how your research is developing. In order to do this, you may need to start by looking for small grants aimed at early career researchers or at specific charities and bodies that offer funding for research in their field, community or geographical location. The competition is likely to be less and your expertise in a narrow area may be appreciated. It is also useful to look to colleagues with experience of successful funding applications to ask for their ideas of where to apply. They may also be willing to review your application prior to its submission and/or to share examples of successful applications. Use your network to maintain links with other researchers so that you can hear about people looking for collaborators and make contact with them. Sign up for newsletters about upcoming funding calls. Most research and knowledge transfer departments in universities will send out collations of such calls on a regular basis. They are also helpful in discussing requirements of different funders and devising the budget being applied for. There are a number of free databases listing research funders and funded researchers (e.g. ScientifyRESEARCH, https://www.scientifyresearch.org/all-funding-opportunities-for-researchers-worldwide, and the Science Foundation Ireland, https://www.sfi.ie/funding/researcher-database). Similarly, make it clear to your supervisor or mentor that you are interested in being considered as a co-researcher on future applications that they may be making. Conversely, if you are planning on making a funding application, then it can be useful to invite other researchers to join you, either to bring different perspectives to the research and/or to bring different methods to those in which you are an expert. Depending on the research topic and funder's requirements, you can put together a multidisciplinary team or one made up of community members and other stakeholders alongside academic researchers.

Enhancing your profile

Prior to the advance of technology and social media, the main way to draw attention to your research was via publications Now, however, there are many platforms in addition to academic journals in which you can disseminate your work and your role as a researcher in it. We will describe some of these below but, first, will consider other ways in which you can enhance your profile as a researcher by using personal skills and characteristics.

Personally

The key personal skill to enhancing your profile as a researcher is in being a good communicator, contributor and supporter of others. Your researcher role will most likely be associated with the university school or department in which you or co-researchers are based, so it is a good start to connect with your immediate colleagues there. You can show interest in their research, perhaps offering to review papers or grant applications, and cultivate interest in yours. Seeking out more experienced researchers in the department (or those who are connected to it externally, such as visiting scholars or examiners) for advice or collaboration raises awareness of your research area and demonstrates that you are a researcher who values their input. You can also offer to deliver departmental seminars and workshops to give more detail and seek feedback on your research. Similarly, attending seminars delivered by other researchers shows support, expands your knowledge and interest and offers ways of providing feedback. Many departments produce regular newsletters, and you can ensure that your successes in acquiring funding or having your research published are included there. Schools and departments sometimes offer seed funding to their members, so you should ensure you are aware of calls for these and apply for them if appropriate or bring it to the attention of others. Successful applications can sometimes mean that you can employ a research assistant, and choosing to employ doctoral or postdoctoral members of your department demonstrates that you are a collegiate and supportive researcher. It can be rewarding as well as beneficial to find ways to support PhD students, whether you are supervising them or not. They can be keen to advance their own knowledge of being a researcher and are always up to date on latest developments in research areas. Taking time to discuss their work with them and responding to any direct approaches they may make to you are mutually beneficial. PhD students who remain in the academe can become valuable co-researchers and writers as you both develop as researchers.

Looking more widely in the university to enhance your profile, you should ensure that you have access to lists sent out to research staff that detail national and international funding calls. When these fall in your area of interest, you can consider a cross-disciplinary or cross-university team – either forming one or joining one. This will not only bring multi-perspective expertise but also broaden your network. Many teams benefit from employing researchers with different research methods, from different research settings and with different knowledge of subject areas. Being part of these teams broadens the ways in which your topic can be understood and enables research questions that you may not have thought of to be asked. This enhances your profile as a researcher who is open to learning from others and can think creatively. You can offer your help in organizing internal conferences and in spreading the word to other researchers.

It is also valuable to be connected to the university's research office. It will be a useful source of research information and getting to know who works in it will mean you can approach them more easily for advice and help with grant applications.

Working with researchers in other universities also shows how your research can be disseminated more widely and bring attention to your work and that of your university. This can lead to opportunities to invite researchers to come and talk about their research and/or visit their institution to talk about yours. Such connections can also lead to requests for you to take up roles such as external examining for PhD vivas.

Finally, consider joining associations and special interest groups relevant to your research. Bodies such as the British Psychological Society not only provide sources of information about your discipline but also comprise many different sections. Taking up membership in sections that are of interest to you means that you can contribute to meetings and bulletins, apply for funding and attend conferences. You can also apply to deliver workshops and seminars. Going a step further and joining committees also provides opportunities to enhance your profile as a researcher as does contributing to supporting, and raising awareness of, the work of others.

Technologically

The range of online platforms now available offers many opportunities for enhancing your profile. It is also an increasing expectation of universities and other institutions that you develop an online profile. A key to success to using these is in ensuring they remain up to date – if you have a website that hasn't been updated, it will look like you are not doing research and visitors will diminish. Maintaining online information sources is an ongoing task and one that you should either schedule in your own timetable or recruit help with. Depending on the platform(s) you are using, you may want to upload/update information and notifications on a weekly basis (e.g. X, formerly known as Twitter), monthly basis (e.g. participation in an online discussion) and/or six-monthly (e.g. LinkedIn). Making these tasks part of your work schedule means that they are more likely to happen.

Using online media for enhancement of your own profile is a two-way process – not only can you inform others of the work you are doing but also learn about (and comment on/congratulate) others on theirs. You will also learn about what is of interest to other researchers in your field, whether it is new publications, funding calls or requests for collaboration. You can seek or give advice in interactive fora such as ResearchGate, and increasingly, researchers are using social media to recruit participants for new research projects. There are many free platforms (some listed below) which also do the work of collating your work for you.

Googlescholar is a useful collation of your publications and has the bonus of updating itself. It is a well-frequented source of information about researchers' works. You will need to register yourself on it with a verifiable institutional email address.

LinkedIn is useful for promoting your work and is often sourced by potential employers. The information you upload to it can range from reposting of interesting threads to publicizing your own recent achievements.

ResearchGate allows you to see new publications as well as promote your own and connect with other researchers for support and collaboration. Its metrics give you information about your citations and who is reading your work.

Open Researcher and Contributor ID (ORCID) is a unique identifier for researchers that aims to connect individuals to their contributions to research, scholarship and innovation across disciplines, borders and time periods. Once you have registered and been assigned your unique 16-digit number, you can use it on publications and grant applications, as well as on your research institution management system to give permission for information exchange between these systems. All your research is linked together, and you can set the level of visibility you wish it to have.

Publons aims to increase the attribution of the work done by peer reviewers. By agreeing to have your name and review included on Publons, you can track, verify and showcase your peer review activities and contributions online.

Many of the above offer low maintenance and workload but there are other ways in which you can enhance your profile if you are willing to be a little more proactive.

Academic blogs are usually written in an informal style and discuss all aspects of being a researcher, from coping with PhD life (e.g. the Thesis Whisperer) to how to develop your career as a researcher (e.g. The Professor is In) to specific issues and challenges in individual disciplines (e.g iRashia). Writing a blog can be relatively quick and straightforward, particularly if it is on an aspect of research you are particularly interested in, but the challenge is to maintain it regularly by adding new posts and/or inviting contributions from guest bloggers. Blogging can offer ways of connecting with a potentially worldwide network of researchers.

Like blog-writing, making a podcast can bring you followers and connections with other researchers. They offer an alternative to reading and can cover a range of topics relevant to researchers, from well-being (e.g. The Self-Compassionate Professor) to career development (e.g. Academic Writing Amplified). Some journal publishers such as Taylor & Francis also offer podcasts for researchers. They range from imagining the skills of researchers of the future to how to write articles for publication. Starting your own podcast or appearing as a guest on existing ones can help to raise your profile as a researcher.

With more effort (and technical skills), you can consider making YouTube videos to talk about and present your work. It will probably be useful if you can use the technology offered by your university, and having their name attached to your video may enhance its visibility.

Finally, consider contacting local or national radio stations and newspapers (including The Conversation, which is a not-for-profit network of media outlets that publish online articles written by academics and researchers) to request that you discuss your work on air or in print. This is a good way to reach the general public and enhances your profile as a researcher who contributes to knowledge transfer outside the academe. Most universities and some associations offer media training to help prepare for this.

Chapter summary

This chapter has described ways in which you are researched as a researcher and how you can use these to enhance your profile. They range from expectations and monitoring of your research activities and outputs to using your skills to proactively connect and communicate with a wider audience. Whilst you may not always feel comfortable doing this, it is increasingly expected that you do so. Support is available and the rewards can be many, from showcasing your research to getting feedback, advice and support recommendations for developing it. You and your research are likely to become increasingly recognized and these activities can help with networking, gaining promotion and seeking employment.

Further reading

Mertler, C. A. (2023). *Disseminating your action research: A practical guide to sharing the results of practitioner research*. Routledge.
This book offers professional educators of any academic level, subject area or position the tools, guidance, techniques and strategies to disseminate, share, publish and promote the results of action research projects and studies.

Morley, J. (2023). *Academic phrasebank: An academic writing resource for students and researchers*. University of Manchester.
Developed from the University of Manchester website, this book makes available many thousands of commonly used phrases found in academic and scientific writing. The phrases are organized and presented according to the major sections and subsections of a university thesis or dissertation.

Women in Academia Support Network. (2022). *ResearchHER: The power and potential for research careers for women*. Emerald Publishing Limited.
This book demonstrates the diversity of scholarship and research, and the women leading the way. It offers an A–Z of research and researchers from around the world, exploring who researchers are and what they really do, all whilst celebrating female scholarship. Each short chapter offers an insight into the background and journey into the research career of its author, what they are currently researching, their top tips for budding researchers and fun facts and activities.

Bibliography

Abdelaal, G. (2020). Coping with imposter syndrome in academia and research. *The Biochemist, 42*(3), 62–64.

Adjerid, I., & Kelley, K. (2018). Big Data in psychology: A framework for research advancement. *American Psychologist, 73*(7), 899.

Álvarez, E. G., Sintas, J. L., & Martínez, A. S. (2012). Challenges and trends in digital leisure: Transformation of dimensions, experiences and business models. *Arbor, 188*(754), 395–407.

American Psychological Association. (2016). *Ethical Principles of Psychologists and Code of Conduct*. Available at http://www.apa.org/ethics/code/index.aspx. Accessed on 24 February 2024.

Anderson, J. (2004). Talking whilst walking: A geographical archaeology of knowledge. *Area, 36*(3), 254–261.

Arnot, M., David, M. E., & Weiner, G. (1996). *Educational reforms and gender equality in schools*. Equal Opportunities Commission.

Bager-Charleson, S. (2019). 'She was on my side, and grounded me when I needed it': Research supervision in the field of therapy, based on counsellors' and psychotherapists' views on their engagement with research. *Counselling and Psychotherapy Research, 19*(4), 358–365.

Bager-Charleson, S., & McBeath, A. G. (2020). *Enjoying research in counselling and psychotherapy*. Springer International Publishing.

Bager-Charleson, S., & McBeath, A. G. (2021). What support do therapists need to do research? A review of studies into how therapists experience research. *Counselling and Psychotherapy Research, 21*(3), 555–569.

Ballamingie, P., & Johnson, S. (2011). The vulnerable researcher: Some unanticipated challenges of doctoral fieldwork. *Qualitative Report, 16*(3), 711–729.

Barkhuizen, N., Rothmann, S., & Van De Vijver, F. J. (2014). Burnout and work engagement of academics in higher education institutions: Effects of dispositional optimism. *Stress and Health, 30*(4), 322–332.

Bateson, G., & Mead, M. (1942). *Balinese character: A photographic analysis*. The New York Academy of Sciences.

Baveye, P. C. (2021). Objectivity of the peer-review process: Enduring myth, reality, and possible remedies. *Learned Publishing, 34*(4), 696–700.

Berger, R. (2021). Studying trauma: Indirect effects on researchers and self-and strategies for addressing them. *European Journal of Trauma & Dissociation, 5*(1), 100149.

Biddle, C., & Schafft, K. A. (2015) Axiology and anomaly in the practice of mixed methods work: Pragmatism, valuation, and the transformative paradigm. *Journal of Mixed Methods Research, 9*(4), 320–334.

Bluvstein, I., Ifrah, K., Lifshitz, R., Markovitz, N., & Shmotkin, D. (2021). Vulnerability and resilience in sensitive research: The case of the quantitative researcher. *Journal of Empirical Research on Human Research Ethics, 16*(4), 396–402.

Bondi, L. (2014). Understanding feelings: Engaging with unconscious communication and embodied knowledge. *Emotion, Space and Society, 10*, 44–54.

Boone, A., Vander Elst, T., Vandenbroeck, S., & Godderis, L. (2022). Burnout profiles among young researchers: A latent profile analysis. *Frontiers in Psychology, 13*, 839728.

Booysen, L. A., Bendl, R., & Pringle, J. K. (Eds.). (2018). *Handbook of research methods in diversity nanagement, equality and inclusion at work*. Edward Elgar Publishing.

Boyd, D., & Crawford, K. (2012). Critical questions for big data: Provocations for a cultural, technological, and scholarly phenomenon. *Information, Communication & Society, 15*(5), 662–679.

Branney, P., Reid, K., Frost, N., Coan, S., Mathieson, A., & Woolhouse, M. (2019). A context-consent meta framework for designing open (qualitative) data studies. *Qualitative Research in Psychology, 16*(3), 483–502.

Branney, P., Strickland, C., Darby, F., White, L., & Jain, S. (2017). Exploring men's experiences of diagnosis and treatment for prostate cancer. In J. Brooks, & N. King (Eds.). *Applied Qualitative Research in Psychology*. Bloomsbury Publishing.

British Educational Research Association. (2024). Ethical Guidelines for Educational Research. Available at https://www.bera.ac.uk/publication/ethical-guidelines-for-educational-research-fifth-edition-2024-online.

Brody, H. (2021). What were we mapping? From the Inuit Land Use and Occupancy Project to the Southern Kalahari. In U. Dieckmann (Ed.), *Mapping the unmappable?: Cartographic explorations with Indigenous people in Africa*. Columbia University Press.

Bryman, A. (2007). Barriers to integrating quantitative and qualitative research. *Journal of Mixed Methods Research, 1*(1), 8–22.

Burr, H., Berthelsen, H., Moncada, S., Nübling, M., Dupret, E., Demiral, Y., …, Pohrt, A. (2019) The third version of the Copenhagen psychosocial questionnaire. *Safety and Health at Work, 10*(4), 482–503.

Carlson, K. D., & Wu, J. (2012). The illusion of statistical control: Control variable practice in management research. *Organizational Research Methods, 15*(3), 413–435.

Charmaz, K. (2014). Grounded theory in global perspective: Reviews by international researchers. *Qualitative Inquiry, 20*(9), 1074–1084.

Chavez, C. (2008). Conceptualizing from the inside: Advantages, complications, and demands on insider positionality, *The Qualitative Report, 13*(3), 474–494.

Chen, E. E., & Wojcik, S. P. (2016). A practical guide to big data research in psychology. *Psychological Methods, 21*, 458–474.

Clance, P. R., & Imes, S. A. (1978). The imposter phenomenon in high achieving women: Dynamics and therapeutic intervention. *Psychotherapy: Theory, Research & Practice, 15*(3), 241.

Clark, U. S., & Hurd, Y. L. (2020). Addressing racism and disparities in the biomedical sciences. *Nature Human Behaviour, 4*(8), 774–777.

Clarke, N. J., Caddick, N., & Frost, N. (2016). Pluralistic data analysis: Theory and practice. In B. Smith & A. C. Sparkes (Eds.), *Routledge handbook of qualitative research in sport and exercise* (pp. 390–403). Routledge.

Coles, J., Astbury, J., Dartnall, E., & Limjerwala, S. (2014). A Qualitative Exploration of Researcher Trauma and Researchers' Responses to Investigating Sexual Violence. *Violence against women, 20*(1), 95–117.

Colohan, M., Tunariu, A., & O'Dell, P. (2012). Lived experience and discursive context: A twin focus. *Qualitative Methods in Psychology Bulletin, 13*(1), 48–57.

Congdon, V. (2015). The 'lone female researcher': isolation and safety upon arrival in the field. *Journal of the Anthropological Society of Oxford, 7*(1), 15–24.

Cornish, F., Breton, N., Moreno-Tabarez, U., Delgado, J., Rua, M., Aikins, A.dG., & Hodgetts, D. (2023). Participatory action research. *Nature Reviews Methods Primers, 3*(1), 34.

Crenshaw, K. W. (1995). Mapping the margins: Intersectionality, identity politics, and violence against women of color. In M.A. Fineman (Ed.), *The public nature of private violence: Women and the discovery of abuse* (pp. 93–118). Routledge.

Daumiller, M., & Dresel, M. (2020). Researchers' achievement goals: Prevalence, structure, and associations with job burnout/engagement and professional learning. *Contemporary Educational Psychology, 61,* 101843.

De La Mare, D. (2024). Self-compassionate professor podcast. Available at https://danielledelamare.com/subversive-self-compassion/. Accessed 10 January 2024.

Delice, F., Rousseau, M., & Feitosa, J. (2019). Advancing teams research: What, when, and how to measure team dynamics over time. *Frontiers in Psychology, 10,* 1324.

Dempsey, M., Foley, S., Frost, N., Murphy, R., Willis, N., Robinson, S., …, McCarthy, J. (2019). Am I lazy, a drama queen or depressed? A pluralistic analysis of participant and researcher data when analysing accounts of depression posted to an Ireland-based website. *Qualitative Research in Psychology, 19*(2), 1–21.

Dewidar, O., Elmestekawy, N., & Welch, V. (2022). Improving equity, diversity, and inclusion in academia. *Research Integrity and Peer Review, 7*(1), 4.

Diebold, F. X. (2021). What's the big idea? 'Big Data' and its origins. *Significance, 18*(1), 36–37.

Dijkstra, S., Kok, G., Ledford, J. G., Sandalova, E., & Stevelink, R. (2018). Possibilities and pitfalls of social media for translational medicine. *Frontiers in Medicine, 5,* 345.

Downey, H., Hamilton, K., & Catterall, M. (2007). Researching vulnerability: What about the researcher? *European Journal of Marketing, 41*(7/8), 734–739.

Edwards, C. W. (2019). Overcoming imposter syndrome and stereotype threat: Reconceptualizing the definition of a scholar. *Taboo: The Journal of Culture and Education, 18*(1), 3.

Edwards, L. H., Holmes, M. H., & Sowa, J. E. (2019). Including women in public affairs departments: Diversity is not enough. *Journal of Public Affairs Education, 25*(2), 163–184.

Egan, G. (1994). *Working the shadow side: A guide to positive behind-the-scenes management.* Jossey-Bass.

Eley, A., Wellington, J., Pitts, S., & Biggs, C. (2012). *Becoming a successful early career researcher.* Routledge.

Emerald, E., & Carpenter, L. (2015). Vulnerability and emotions in research: Risks, dilemmas, and doubts. *Qualitative Inquiry, 21*(8), 741–750.

Equality and Human Rights Commission. (2021). Available at https://www.equalityhumanrights.com/en/equality-act/protected-characteristics. Accessed on 22 September 2023.

European Commission. The EU data protection reform and Big Data: Factsheet 2016. Available at https://publications.europa.eu/en/publication-detail/-/publication/51fc3ba6-e601-11e7-9749-01aa75ed71a1. Accessed on 19 October 2023.

Favaretto, M., De Clercq, E., Schneble, C. O., & Elger, B. S. (2020). What is your definition of Big Data? Researchers' understanding of the phenomenon of the decade. *PloS One, 15*(2), e0228987.

Finlay, L. (2020). How to write a journal article: Top tips for the novice writer. *European Journal for Qualitative Research in Psychotherapy, 10,* 28–40.

Fisher, C. B. (2000). Relational ethics in psychological research: One feminist's journey. In M. M. Brabeck (Ed.), Practicing feminist ethics in psychology (pp. 125–142). American Psychological Association. https://doi.org/10.1037/10343-006

Fittler, A., Bősze, G., & Botz, L. (2013). Evaluating aspects of online medication safety in long-term follow-up of 136 Internet pharmacies: Illegal rogue online pharmacies flourish and are long-lived. *Journal of Medical Internet Research, 15*(9), e2606.

Flick, U. (2017). *The Sage handbook of qualitative data collection.* Sage Publications Ltd.

Forester, M., Kahn, J. H., & Hesson-McInnis, M. S. (2004). Factor structures of three measures of research self-efficacy. *Journal of Career Assessment, 12*(1), 3–16.

Fossli, G., & Michaelsen, H. C. (2017). When the supervision process falters and breaks down: Pathways to repair. *Supervision of Family Therapy and Systemic Practice* (pp. 227–260). Springer.

Fox, M., Martin, P., & Green, G. (2007). *Doing practitioner research.* Sage Publications Ltd.

Frost, N. (2016). *Practising research: Why you're always part of the research process even when you think you're not.* Palgrave Macmillan.

Frost, N. (2021). *Qualitative research methods in psychology: Combining core approaches*, (2nd ed.). McGraw-Hill Education.

Frost, N., & Holt, A. (2014). Mother, researcher, feminist, woman: Reflections on 'maternal status' as a researcher identity. *Qualitative Research Journal, 14*(2), 90–102.

Gaillard, S., van Viegen, T., Veldsman, M., Stefan, M. I., & Cheplygina, V. (2022). Ten simple rules for failing successfully in academia. *PLoS Computational Biology, 18*(12), e1010538.

Ganga, D., & Scott, S. (2006). Cultural 'insiders' and the issue of positionality in qualitative migration research: Moving 'across' and moving 'along' researcher-participant divides. *Forum Qualitative Sozialforschung/Forum: Qualitative Social Research, 7*(3). https://doi.org/10.17169/fqs-7.3.13.

Gardner, H. (1983). *Frames of mind.* Basic Books.

Gelman, A., & Hennig, C. (2017). Beyond subjective and objective in statistics. *Journal of the Royal Statistical Society Series A: Statistics in Society, 180*(4), 967–1033.

Gewin, V. (2021). Pandemic burnout is rampant in academia. *Nature, 591*(7850), 489–492.

Giampapa, F. (2011). The politics of 'being and becoming' a researcher: Identity, power, and negotiating the field. *Journal of Language, Identity & Education, 10*(3), 132–144.

Giles, J. (2005). Researchers break the rules in frustration at review boards. *Nature, 438*(7065), 136–138.

Gilligan, C. (1977). In a different voice: Women's conception of the self and of morality. *Harvard Educational Review, 47*(4), 481–517. https://doi.org/10.17763/haer.47.4.g6167429416hg5l0

Gilligan, C. (1993). *In a different voice: Psychological theory and women's development.* Harvard University Press.

Goleman, D. (1998). *Working with Emotional Intelligence.* Bantam.

Grandin, T. (1992). An inside view of autism. In E. Schopler & G. B. Mesibov (Eds.), *High-functioning individuals with autism* (pp. 105–126). Springer US.

Guidetti, G., Converso, D., Di Fiore, T., & Viotti, S. (2022). Cynicism and dedication to work in post-docs: Relationships between individual job insecurity, job insecurity climate, and supervisor support. *European Journal of Higher Education, 12*(2), 134–152.

Gustavii, B. (2012). *How to prepare a scientific doctoral dissertation based on research articles.* Cambridge University Press.

Hansen, S., & Flynn, D. A. (2015). Longitudinal photo-documentation: Recording living walls. *Street Art & Urban Creativity Journal, 1*(1), 26–31.

Harding, S. (1995). 'Strong objectivity': A response to the new objectivity question. *Synthese, 104*, 331–349.

Harding, S. (2013). Rethinking standpoint epistemology: What is 'strong objectivity'? In L. Alcoff (Ed.), *Feminist epistemologies* (pp. 49–82). Routledge.

Haven, T. L., de Goede, M. E. E., Tijdink, J. K., & Oort, F. J. (2019). Personally perceived publication pressure: Revising the Publication Pressure Questionnaire (PPQ) by using work stress models. *Research Integrity and Peer Review, 4*(1): 1–9.

Hazell, C. M., Chapman, L., Valeix, S. F., Roberts, P., Niven, J. E., & Berry, C. (2020). Understanding the mental health of doctoral researchers: A mixed methods systematic review with meta-analysis and meta-synthesis. *Systematic Reviews, 9*(1): 1–30.

Hein, J. R., Evans, J., & Jones, P. (2008) Mobile methodologies: Theory, technology and practice. *Geography Compass, 2*(5), 1266–1285.

Heinrich, E., Hill, G., Kelder, J. A., & Picard, M. (2024). Group-based journal review opportunities for researcher development and enjoyment. *International Journal for Academic Development*, 1–16.

Hellawell, D. (2006). Inside–out: Analysis of the insider–outsider concept as a heuristic device to develop reflexivity in students doing qualitative research. *Teaching in Higher Education, 11*(4), 483–494.

Hesse-Biber, S. (2010). Qualitative approaches to mixed methods practice. *Qualitative Inquiry, 16*(6), 455–468.

Hitchcock, J. H., & Onwuegbuzie, A. J. (Eds.). (2022). *The Routledge handbook for advancing integration in mixed methods research*. Routledge.

Hochschild, A. R. (2003). *The commercialization of intimate life: Notes from home and work*. University of California Press.

Holmes, A. G. D. (2020). Researcher positionality - A consideration of its influence and place in qualitative research: A new researcher guide. *Shanlax International Journal of Education, 8*(4), 1–10.

Holt, A. (2010). Using the telephone for narrative interviewing: A research note. *Qualitative Research, 10*(1), 113–121.

Hubbard, G., Backett-Milburn, K., & Kemmer, D. (2001). Working with emotion: Issues for the researcher in fieldwork and teamwork. *International Journal of Social Research Methodology, 4*(2), 119–137.

Ibrahim, A. M., Lillemoe, K. D., Klingensmith, M. E., & Dimick, J. B. (2017). Visual abstracts to disseminate research on social media: A prospective, case-control crossover study. *Annals of Surgery, 266*(6), e46–e48.

Jacknis, I. (1988). Margaret Mead and Gregory Bateson in Bali: Their use of photography and film. *Cultural Anthropology, 3*(2), 160–177.

Jamieson, M. K., Govaart, G. H., & Pownall, M. (2023). Reflexivity in quantitative research: A rationale and beginner's guide. *Social and Personality Psychology Compass, 17*(4), e12735.

Jones, B. F. (2021). The rise of research teams: Benefits and costs in economics, *Journal of Economic Perspectives, 35*(2), 191–216.

Jones, H., & Stanley, G. (2010). Collaborative action research: A democratic undertaking or a web of collusion and compliance? *International Journal of Research & Method in Education, 33*(2), 151–163.

Kaatz, A., Gutierrez, B., & Carnes, M. (2014). Threats to objectivity in peer review: The case of gender. *Trends in Pharmacological Sciences, 35*(8), 371–373.

Kaisler, S., Armour, F., Espinosa, J. A., & Money, W. (2013). Big data: Issues and challenges moving forward. *46th Hawaii International Conference on System Sciences* (pp. 995–1004).

Kalpazidou, E., Schmidt, E., Ovseiko, P.V., Henderson, L.R., & Kiparoglou, V. (2020) Understanding the Athena SWAN award scheme for gender equality as a complex social intervention in a complex system: Analysis of Silver award action plans in a comparative European perspective. *Health Research Policy and Systems 18*(19). https://doi.org/10.1186/s12961-020-0527-x

Kellogg, W. K. (2004). Using logic models to bring together planning, evaluation, and action: Logic model development guide. WK Kellogg Foundation.

Kelly, C., Dansereau, L., Sebring, J., Aubrecht, K., FitzGerald, M., Lee, Y., Williams, A., & Hamilton-Hinch, B. (2022). Intersectionality, health equity, and EDI: What's the difference for health researchers? *International Journal for Equity in Health, 21*(1), 182.

Kerstetter, K. (2012). Insider, outsider, or somewhere between: The impact of researchers' identities on the community-based research process. *Journal of Rural Social Sciences, 27*(2), 7.

King, J. (2023). Indigeneity, positionality, and ethical space: Navigating the in-between of Indigenous and settler academic discourse. *Papers on Postsecondary Learning and Teaching, 6*, 36–48.

King, N., Finlay, L., Ashworth, P., Smith, J. A., Langdridge, D., & Butt, T. (2008). 'Can't really trust that, so what can I trust?': A polyvocal, qualitative analysis of the psychology of mistrust. *Qualitative Research in Psychology, 5*(2), 80–102.

Kitchin, R., & McArdle, G. (2016) What makes Big Data, Big Data? Exploring the ontological characteristics of 26 datasets. *Big Data & Society, 3*(1), 2053951716631130.

Knapp, S., Gottlieb, M. C., & Handelsman, M. M. (2018). The benefits of adopting a positive perspective in ethics education. *Training and Education in Professional Psychology, 12*(3), 196.

Knapp, S. J., & VandeCreek, L. D. (2006). Ethical decision making. In S. J. Knapp & L. D. VandeCreek, *Practical ethics for psychologists: A positive approach* (pp. 39–49). American Psychological Association. https://doi.org/10.1037/11331-004

Knapp, S. J., Van de Creek, L. D. & Fingerhut, R. (2017). *Practical ethics for psychologists: A positive approach.* American Psychological Association.

Koehne, B., Shih, P. C., & Olson, J. S. (2012). Remote and alone: Coping with being the remote member on the team. *Proceedings of the ACM 2012 Conference on Computer Supported Cooperative Work* (pp. 1257–1266).

Kozinets, R. V. (1999). E-tribalized marketing?: The strategic implications of virtual communities of consumption. *European Management Journal, 17*(3), 252–264.

Kozinets, R. V. (2002). The field behind the screen: Using netnography for marketing research in online communities. *Journal of Marketing Research, 39*(1), 61–72.

Kozinets, R. V. (2015). *Netnography: Redefined.* Sage Publications Ltd.

Kozinets, R. V. (2019). *Netnography: The essential guide to qualitative social media research*, (3rd ed.). Sage Publications Ltd.

Lajoie, C., Poleksic, J., Bracken-Roche, D., MacDonald, M. E., & Racine, E. (2020). The concept of vulnerability in mental health research: A mixed methods study on researcher perspectives. *Journal of Empirical Research on Human Research Ethics, 15*(3), 128–142.

Landers, R. N., Brusso, R. C., Cavanaugh, K. J., & Collmus, A. B. (2016). A primer on theory-driven web scraping: Automatic extraction of Big Data from the internet for use in psychological research. *Psychological Methods, 21*, 475–492.

Langer, R., & Beckman, S. C. (2005) Sensitive research topics: Netnography revisited. *Qualitative Market Research: An International Journal, 8*(2), 189–203.

Lee, J., & Ingold, T. (2020). Fieldwork on foot: Perceiving, routing, socializing. In S. Coleman & P. Collins (Eds.), *Locating the field: Space, place and context in anthropology* (pp. 67–85).Routledge.

Lee, R. M. (1993). *Doing research on sensitive topics.* Sage Publications Ltd.

Lee, Y. N., Walsh, J. P., & Wang, J. (2015). Creativity in scientific teams: Unpacking novelty and impact. *Research Policy, 44*(3), 684–697.

Levecque, K., Anseel, F., De Beuckelaer, A., Van der Heyden, J., & Gisle, L. (2017). Work organization and mental health problems in PhD students. *Research Policy, 46*(4). 868–879.

Loxley, A., & Seery, A. (2008). Some philosophical and other related issues of insider research. In P. Sikes & A. Potts (Eds.), *Researching education from the inside: Investigations from within* (pp. 23–40). Routledge.

Luca, J., & Tarricone, P. (2001). Does emotional intelligence affect successful teamwork? Available at *Research Online*. https://ro.ecu.edu.au/ecuworks/4834.

Lupton, D. (2015). The thirteen Ps of Big Data. *This Sociological Life* blog. https://simplysociology.wordpress.com/2015/05/11/the-thirteen-ps-of-big-data/. Accessed on 31 January 2023.

Majundar, A. (2024). A pluralistic narrative exploration of spiritual journeys and experiences of psychological therapists with a longstanding and ongoing meditation practice, unpublished thesis. City University of London.

Mann, S. (2019). *Why do I feel like an imposter?: How to understand and cope with imposter syndrome.* Watkins Media Limited.

Mantai, L. (2017). Feeling like a researcher: Experiences of early doctoral students in Australia. *Studies in Higher Education, 42*(4), 636–650.

Maslach, C., Schaufeli, W. B., & Leiter, M. P. (2001). Job burnout. *Annual Review of Psychology, 52*(1), 397–422.

Mavhandu-Mudzusi, A. H. (2023). Application of ethics in the South African rural context. *International Journal of Qualitative Methods, 22*, 16094069231193592.

Mayer, J. D., & Salovey, P. (2007). *Mayer-Salovey-Caruso emotional intelligence test.* Multi-Health Systems Incorporated.

Mayer-Schönberger, V., & Cukier, K. (2013). *Big Data: A revolution that will transform how we live, work, and think.* Houghton Mifflin Harcourt.

McAdams, D. P. (1993). *Stories we live by: Personal myths and the making of the self.* Morrow.

McKay, S., Veale, A., Worthen, M. E., & Wessells, M. (2011). Building meaningful participation in (re) integration among war-affected young mothers in Liberia, Sierra Leone and Northern Uganda. *Intervention: International Journal of Mental Health, Psychosocial Work and Counselling in Areas of Armed Conflict, 9*(2), 108.

Mellors-Bourne, R., & Metcalfe, J. (2017). *Five steps forward: Progress in implementing the concordat to support the career development of researchers 2008–2017.* Vitae.

Menard, C. B., & Shinton, S. (2022). The career paths of researchers in long-term employment on short-term contracts. Case study from a UK university. *Plos One, 17*(9), e0274486.

Miller, A. N., Taylor, S. G., & Bedeian, A. G. (2011). Publish or perish: Academic life as management faculty live it. *Career Development International, 16*(5), 422–445.

Miller, T. (2015). Going back: 'Stalking', talking and researcher responsibilities in qualitative longitudinal research. *International Journal of Social Research Methodology, 18*(3), 293–305.

Mockler, N. (2014). When 'research ethics' become 'everyday ethics': The intersection of inquiry and practice in practitioner research. *Educational Action Research, 22*(2), 146–158.

Mohammed, S., Hamilton, K., & Lim, A. (2008). The incorporation of time in team research: Past, current, and future. In E. Salas, G. F. Goodwin & C. Shawn Burke (Eds.), *Team effectiveness in complex organizations* (pp. 355–382). Taylor & Francis Group.

Moncur, W. (2013). The emotional wellbeing of researchers: Considerations for practice. *Proceedings of the SIGCHI conference on human factors in computing systems* (pp. 1883–1890).

Morse, J. M., Cheek, J., & Clark, L. (2018). Data-related issues in qualitatively driven mixed-method designs: Sampling, pacing, and reflexivity. In U. Flick (Ed.), *The SAGE handbook of qualitative data collection* (pp. 564–584). Sage Publications Ltd.

Newton, B. J. (2021). Creating cultural safety as an Aboriginal teacher in a class of non-Aboriginal university students. *Australian Social Work, 74*(1), 4–12.

Nygaard, L. P., & Solli, K. (2020). *Strategies for writing a thesis by publication in the social sciences and humanities.* Routledge.

Onwuegbuzie, A. J., & Leech, N. L. (2005). On becoming a pragmatic researcher: The importance of combining quantitative and qualitative research methodologies. *International Journal of Social Research Methodology, 8*(5), 375–387.

Paltridge, B., & Starfield, S. (2023). The PhD by publication in the humanities and social sciences: a cross country analysis. *Journal of Further and Higher Education, 47*(7), 863–874.

Parker-Jenkins, M. (2018). Mind the gap: Developing the roles, expectations and boundaries in the doctoral supervisor–supervisee relationship. *Studies in Higher Education, 43*(1), 57–71.

Parpart, J. L. (2010). Choosing silence: Rethinking voice, agency and women's empowerment. In R. Ryan-Flood & R. Gill (Eds.), *Secrecy and silence in the research process: Feminist reflections.* Routledge.

Partington, D. (2009). *Essential skills for management research.* Sage Publications.

Pelletier, C. A., Pousette, A., Ward, K., & Fox, G. (2020). Exploring the perspectives of community members as research partners in rural and remote areas. *Research Involvement and Engagement, 6*, 1–10.

Pezalla, A. E., Pettigrew, J., & Miller-Day, M. (2012). Researching the researcher-as-instrument: An exercise in interviewer self-reflexivity, *Qualitative Research, 12*(2), 165–185.

Pinto, R. M., Spector, A. Y., & Rahman, R. (2019). Nurturing practitioner-researcher partnerships to improve adoption and delivery of research-based social and public health services worldwide. *International Journal of Environmental Research and Public Health, 16*(5), 862.

Pinto, R. M., Spector, A. Y., Rahman, R., & Gastolomendo, J. D. (2015). Research advisory board members' contributions and expectations in the USA. *Health Promotion International, 30*(2), 328–338.

Podschuweit, N. (2021). How ethical challenges of covert observations can be met in practice. *Research Ethics, 17*(3), 309–327.

Purdy, E., Symon, B., Marks, R. E., Speirs, C., & Brazil, V. (2023). Exploring equity, diversity, and inclusion in a simulation program using the SIM-EDI tool: The impact of a reflexive tool for simulation educators. *Advances in Simulation, 8*(1), 11.

Roberts, S. O., Bareket-Shavit, C., Dollins, F. A., Goldie, P. D., & Mortenson, E. (2020). Racial inequality in psychological research: Trends of the past and recommendations for the future. *Perspectives on Psychological Science, 15*(6), 1295–1309.

Ryan, L., & Golden, A. (2006). 'Tick the box please': A reflexive approach to doing quantitative social research. *Sociology, 40*(6), 1191–1200.

Ryan-Flood, R., & Gill, R. (2010). *Secrecy and silence in the research process: Feminist reflections.* Routledge.

Saarijärvi, M., & Bratt, E-L. (2021).When face-to-face interviews are not possible: Tips and tricks for video, telephone, online chat, and email interviews in qualitative research. *European Journal of Cardiovascular Nursing, 20*(4), 392–396.

Salas, E., Cooke, N. J., & Rosen, M. A. (2008). On teams, teamwork, and team performance: Discoveries and developments. *Human Factors, 50*(3), 540–547.

Savin-Baden, M., & Major, C. (2023). *Qualitative research: The essential guide to theory and practice.* Routledge.

Scott, S., Hinton-Smith, T., Härmä, V., & Broome, K. (2012). The reluctant researcher: Shyness in the field. *Qualitative Research, 12*(6), 715–734.

Shankman, P. (2009). *The trashing of Margaret Mead: Anatomy of an anthropological controversy.* University of Wisconsin Press.

Sheller, M., & Urry, J. (2006). The new mobilities paradigm. *Environment and Planning, 38,* 207–226.

Sherry, E. (2013). The vulnerable researcher: Facing the challenges of sensitive research. *Qualitative Research Journal, 13*(3), 278–288.

Sikic Micanovic, L., Stelko, S., & Sakic, S. (2019). Who else needs protection? Reflecting on researcher vulnerability in sensitive research. *Societies, 10*(1), 3.

Slatin, C., Galizzi, M., Melillo, K. D., et al. (2004). Conducting interdisciplinary research to promote healthy and safe employment in health care: Promises and pitfalls. *Public Health Reports, 119*(1), 60–72.

Smith, C., & Ulus, E. (2020). Who cares for academics? We need to talk about emotional well-being including what we avoid and intellectualise through macro-discourses. *Organization, 27*(6), 840–857.

Smoliga, J. M., & Kendall, C. J. (2017). Axe science hype from social media. *Nature, 54*(2), 31.

Soja, E., & Thirdspace, W. (1996). *Journeys to Los Angeles and other real-and-imagined places.* Malden, MA: Blackwell.

Steele, C. M., & Aronson, J. (1995). Attitudes and social cognition. *Journal of Personality and Social Psychology, 69*(5), 797–811.

Stefan, M. (2010). A CV of failures. *Nature, 468,* 467.

Stronach, I., & Morris, B. (1994). Polemical notes on educational evaluation in the age of 'policy hysteria'. *Evaluation & Research in Education, 8*(1–2), 5–19.

Templeton, R. (2019). The effects of unsupportive supervision on doctorate completions. In T. M. Machin, M. Clarà & P. A. Danaher (Eds.), *Traversing the doctorate: Reflections and strategies from students, supervisors and administrators* (pp. 303–317). Palgrave Macmillan.

Thrift, N. (2016). Understanding the affective spaces of political performance1. In M. Smith, J. Davidson & L. Bondi (Eds.), *Emotion, place and culture* (pp. 79–95). Routledge.

Thurairajah, K. (2019). Uncloaking the researcher: Boundaries in qualitative research. *Qualitative Sociology Review,* (1), 132–147.

Todd, J. D. (2021). Experiencing and embodying anxiety in spaces of academia and social research. *Gender, Place & Culture, 28*(4), 475–496.

Uzzi, B., Mukherjee, S., Stringer, M., & Jones, B. (2013). Atypical combinations and scientific impact. *Science, 342*(6157), 468–472.

Valentino, A. L., & Juanico, J. F. (2020). Overcoming barriers to applied research: A guide for practitioners. *Behavior Analysis in Practice, 13*(4), 894–904.

Vander Elst, T., De Witte, H., & De Cuyper, N. (2013). The Job Insecurity Scale: A psychometric evaluation across five European countries. *European Journal of Work and Organizational Psychology, 23*(3), 364–380.

van Zyl, I., & Sabiescu, A. (2020). Toward intersubjective ethics in community-based research. *Community Development, 51*(4), 303–322.

Veale, A., McKay, S., Worthen, M., & Wessells, M. (2013). Participation as principle and tool in social reintegration: Young mothers formerly associated with armed groups in Sierra Leone, Liberia, and Northern Uganda. *Journal of Aggression, Maltreatment & Trauma, 22*(8), 829–848.

Warr, D. J. (2004). Stories in the flesh and voices in the head: Reflections on the context and impact of research with disadvantaged populations. *Qualitative Health Research, 14*(4), 578–587.

Warren, J., & Marz, N. (2015). *Big Data: Principles and best practices of scalable real-time data systems.* Manning Publications.

Wengraf, T. (2004). *The Biographic-Narrative Interpretive Method,* short guide.

Williamson, P. (2016). Take the time and effort to correct misinformation, *Nature,* *540,* 171.

Willig, C. (2021). *Introducing qualitative research in psychology,* 4th ed. McGraw Hill Education.

Woodby, L. L., Williams, B. R., Wittich, A. R., & Burgio, K. L. (2011). Expanding the notion of researcher distress: The cumulative effects of coding. *Qualitative Health Research, 21*(6), 830–838.

Woolston, C. (2020). Pandemic darkens postdocs' work and career hopes. *Nature,* *585*(7824), 309–312.

Xuan, J., & Ocone, R. (2022). The equality, diversity and inclusion in energy and AI: Call for actions. *Energy and AI, 8,* 100152.

Yakushko, O., Badiee, M., Mallory, A., & Wang, S. (2011). Insider outsider: Reflections on working with one's own communities. *Women & Therapy, 34*(3), 279–292.

Index

Printed and bound by CPI Group (UK) Ltd, Croydon, CR0 4YY

27/03/2025

01837455-0011